Island Tourism Policy and Sustainable Development

This insightful and timely book is the first of its kind to explore specific policies, issues, challenges, and practices that will enhance the sustainable development of tourism in island destinations, including island nations, twin-island nations, and sub-national island jurisdictions (SNIJs).

Islands are faced with a myriad of challenges: economic failure, natural disasters, political upheavals, and socio-cultural dilemmas. Tourism is the most likely means for economic development in many islands and yet, specific tailor-made policies for an island context have received limited exploration and discussion. The policies explored in this volume include those relating to management, marketing, governance, and sustainable development of the tourism sector in islands. This book is 'go-to' guide on the topic and the case studies and best practices throughout the book provide practical knowledge and insight. The volume posits a concise and logically structured review of island tourism in a post-pandemic context, exploring specific tourism policies that will contribute to the enhancement of sustainable tourism development in islands, particularly those in developing countries.

This significant book offers insight into best practices and will be of interest to academics, researchers, policymakers, and students of tourism policy, planning, and sustainable development.

Michelle T. McLeod is a Senior Lecturer at The University of the West Indies (The UWI), Mona Campus, Jamaica. Her tourism industry experience spans over 30 years. Her research interests include tourism development, island tourism, knowledge networks, policy networks, and service productivity. McLeod's three co-edited books are *Knowledge Networks and Tourism*, *Tourism Management in Warm-Water Island Destinations*, and *Island Tourism Sustainability and Resiliency*. McLeod has served on The UWI COVID-19 Task Force as the Tourism Expert.

Routledge Advances in Tourism
Edited by **Stephen Page**, *Hertfordshire Business School,*
University of Hertfordshire, UK

Hosting the Olympic Games
Uncertainty, Debates, and Controversy
Edited by Marie Delaplace and Pierre-Olaf Schut

Encountering Nazi Tourism Sites
Derek Dalton

Evaluating the Local Impacts of the Rio Olympics
Edited by Marcelo Neri

Mountain Resort Marketing and Management
Armelle Solelhac

Ageing and the Visitor Economy
Global Challenges and Opportunities
Stephen J. Page and Joanne Connell

Asian Mobilities Consumption in a Changing Arctic
Edited by Young-Sook Lee

Tourism and Foreign Direct Investment
Issues, Challenges and Prospects
Edited by H. Cristina Jönsson

Chinese Outbound Tourist Behaviour
An International Perspective
Edited by Jun Wen and Metin Kozak

Island Tourism Policy and Sustainable Development
Michelle T. McLeod

For more information about this series, please visit: www.routledge.com/
advances-in-tourism/book-series/SE0538

Island Tourism Policy and Sustainable Development

Michelle T. McLeod

Routledge
Taylor & Francis Group

LONDON AND NEW YORK

First published 2025
by Routledge
4 Park Square, Milton Park, Abingdon, Oxon OX14 4RN

and by Routledge
605 Third Avenue, New York, NY 10158

Routledge is an imprint of the Taylor & Francis Group, an informa business

© 2025 Michelle T. McLeod

British Library Cataloguing-in-Publication Data
A catalogue record for this book is available from the British Library

ISBN: 978-1-032-56359-6 (hbk)
ISBN: 978-1-032-56360-2 (pbk)
ISBN: 978-1-003-43511-2 (ebk)

DOI: 10.4324/9781003435112

Typeset in Times New Roman
by Newgen Publishing UK

To my parents

Wilfred St. Patrick Allen* and Carmen Sylvia Allen

Born in

St. Vincent and the Grenadines

Contents

PART III
Island tourism planning for sustainable development   **105**

Illustrations

Figures

Tables

Mini case studies

Best practices

Preface

Sustainable development is a critical concept for islands that have the greatest challenge for remaining in existence. The aim of *Island Tourism Policy and Sustainable Development* is to address the gaps when considering sustainable development in island tourism. Island tourism as a field requires separate considerations when it comes to sustainability and resiliency. Islands are primarily tourism destinations but face many challenges in maintaining successful tourism and hospitality industries. One of the greatest challenges is the impact of climatic events such as natural disasters and the shortage of fresh water. Given the range of challenges that islands face as tourism destinations, it was difficult to understand how stakeholders in the tourism sector could find a path towards sustainable development. This volume addresses this gap by outlining the policy, governance, and planning needed for the sustainable development and management of island tourism as a special form of tourism.

This book considers island tourism case studies and best practices from across the globe. The intention was to seek to cover islands with various characteristics in terms of geographic location, history, culture, and economic circumstances. The chapters in the first part of the book have mini cases that provide a starting point to broaden the discussion around tourism policy development in demand and supply of tourism products and services. Given that islands have resource constraints, specific policies that address sustainability and capacity building for policy processes were also included as chapters. The second part of the book explores island tourism governance in detail. Governance was considered in terms of the characteristics of the stakeholders, governance systems, and strategies that are applicable to an island context. A chapter about consensus building and coordination was included based on the external influences many islands experience in governing the tourism sector. The third part is a practical application of sustainable development to island tourism. This part includes best practices about the islands of Tahiti sustainable tourism planning in Chapter 10, and the Commonwealth of Dominica plans to become the world's first climate resilient nation in Chapter 13. The third part about Island Tourism Planning for Sustainable Development includes a *Research in Focus* in Chapters 11, 12, and 13, and includes elements of a research study about sustainability indicators in several island case studies, to map a path forward for sustainable development in island tourism.

This book comes at a time to support the evolution of island tourism as a separate field in the tourism academy. Islands are 'natural' tourism destinations based on the aesthetic appeal and attractiveness of island features. Research into the sustainability of island tourism destinations has been lagging with scholars addressing the field of island tourism studies on a case-by-case basis. This book contributes to a wholistic view of the sustainable development and management of island tourism by drawing on the author's indigenous experience of over 34 years as a professional and academic in the field of island tourism and drawing from wide as well as disparate sources of the literature surrounding island tourism. The author has embarked on writing this volume with the hope that the field of island tourism will develop further and that island tourism stakeholders will obtain greater practical knowledge, insights, and solutions for solving complex and complicated issues in island tourism.

Acknowledgements

The author appreciates the assistance of her daughter, Miss Abigail McLeod, and the ongoing prompts to complete this book. The author expresses thanks to The University of the West Indies for its research support. The author acknowledges Professor Robertico Croes as Sponsor of a Visiting Research Fellowship at the Dick Pope Snr. Institute for Tourism Studies, at Rosen College of Hospitality Management, University of Central Florida. The author also acknowledges Island Innovation as an information resource about tourism developments in islands.

1 Island tourism impacts and sustainable development

1.1 Introduction

Islands have rich natural and exotic socio-cultural resources, which are contained within a geographical area surrounded by water and designated as islands. Naturally, islands are aligned with the interests of visitors and therefore islands have become one of the most popular tourist destinations. Harrison (2001) has argued that the popularity of islands as tourist destinations is connected to the imagery developed in tourist information. Along with the popularity of islands as tourist destinations, comes the challenges of dealing with influxes of tourists who are temporary visitors. Community members and tourists converge on islands that can be small spaces and such a circumstance results in negative perceptions of tourism (Moyle, Croy, & Weiler, 2010; Sánchez-Cañizares & Castillo-Canalejo, 2014). The temporality of tourism does not mean that there are no impacts. Island temporality has taken on special meanings in island landscapes with a suggestion of multiple temporalities as social impacts (Oroz, 2022) and sandscapes temporalities as physical impacts (Kothari & Arnall, 2020). Impacts of tourism development on islands have been well documented (Carlsen & Butler, 2011; McLeod, Dodds, & Butler, 2022). The scope of this book is to explore the sustainable development of islands, with cases particularly taken from those islands that are described as warm water tourism destinations. Warm water island destinations are those between the Tropic of Cancer and the Tropic of Capricorn (McLeod & Croes, 2018). Island destinations with warm waters have year-round tourism, which brings about challenges in relation to island sustainability.

Sustainable development must be viewed from the perspective of the level of development on island. Resource constraints require prioritisation of actions. Island resource needs to sustain tourism as an economic growth activity, include human resources that are well educated and trained to meet the needs of the industry, and financial capital to create a built environment and infrastructure to support tourism development. Sheldon (2005) has pointed out the special challenges islands face based on the limited resources to develop viable industries, and has recommended greater accessibility and transportation, empowerment, environmental management, knowledge and information systems, marketing and market diversification, stakeholder-involved planning, and visitor management. Resource challenges to

DOI: 10.4324/9781003435112-1

improve the sustainability of islands as tourist destinations must consider several aspects. First, tourism development competes with other developmental activities on islands (Hampton & Christensen, 2007). Second, restrictions in terms of land area and terrain, island accessibility, and the state of the marine environment, limit the capacity of island destinations to support tourism (Buckley, 2002). Third, the size of an island's population makes certain economic activities unprofitable (Eriksen, 2020). Fourth, small islands that are countries often face a paradox of having limits to growth as suggested by Croes (2011). Bearing these challenges in mind, a sustainable development path for islands becomes fuzzy and clarity is needed as to the steps towards sustainable development.

1.2 Island typologies and tourism characteristics

A range of characteristics has been used to classify islands into various types. The most common characteristics that define small islands are population size, land area, and gross domestic product (Croes, 2006). For the purposes of this book, other characteristics that scope islands include the presence of a government, whether central or sub-national governmental body, a population of at least 1,000 persons, and some air or sea infrastructure that connects the island with a mainland. Based on the ease of access many islands have been colonized. Islands in the Caribbean archipelago share Dutch, English, French, and Spanish heritage that makes the region rich in diversity. The societies that develop on an island depends on the intermixing of indigenous people and migrants (Mintz, 1965; Petersen, 2009). The origins of family names may provide a basis to understand the original homeland of island residents, and the 'belongingness' within an island (Cohen, 1995). Cultural diversity added to the natural diversity of islands make these formations an interesting study about sustainable development.

Sustainable tourism development must be framed within the natural characteristics of islands. Island geographies include islands that are volcanic, limestone based with a variety of marine ecosystems. According the National Geographic Society, islands may be categorised into six types: (1) continental islands were once connected to a continent; (2) tidal islands are connected to a mainland and become an island as a result of tidal currents; (3) barrier islands are parallel to coastlines and may develop from coral; (4) oceanic islands are volcanic islands; (5) coral islands are formed from corals; and (6) artificial islands are created in a variety of ways by people (National Geographic, 2023). With the diversity of flora and fauna in a relatively small area, the popularity of islands as tourist destinations has increased. Islands are not only rich in physical features, but can be understood from a human geography perspective (Stratford, 2016).

Island destinations continue to benefit from tourism activities worldwide with The Bahamas and Jamaica showing a larger share of growth than other island countries (Figure 1.1). Barbados, Fiji, and the Maldives have similar levels of growth over 25 years from 1995 to 2020. Samoa has the least number of visitors and the Seychelles' growth in terms of visitors' arrivals, has been consistent although

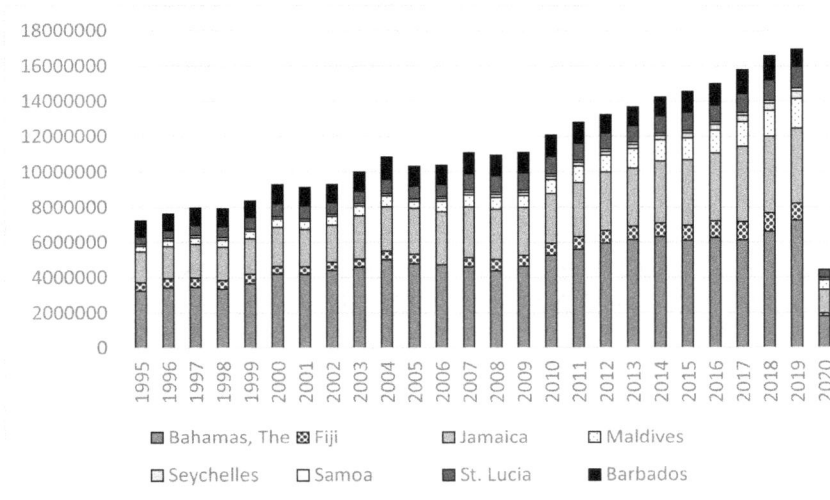

Figure 1.1 International tourist arrivals (1995–2020).

Source: World Development Indicators (World Bank, 2023).

limited. International economic effects in major markets affect tourism demand as visitor arrivals over the period 2007–2009 levelled off, as a result of the 2008 economic crisis in the United States (Figure 1.1). Nonetheless, from 2010, tourist arrivals to the selected islands have a consistent upward thrust with a fall off during the global pandemic in 2020 (Figure 1.1).

1.3 Island tourism impacts

A tourism impact is a long-term change that occurs because of tourism development. The basis for exploring impacts in relation to island tourism has been built on matters surrounding the economic, socio-cultural, and environmental changes that occur in islands (Eslami, Khalifah, Mardani, & Streimikiene, 2018; Sealey, McDonough, & Lunz, 2014). Mason (2020) has noted the factors affecting tourism impacts based on the activities of tourists, location, scale, infrastructure, Tourist Area Life Cycle (Butler, 1980), season, and socio-economic and political contexts. Impacts are counteracted and mitigated using the principles of sustainable tourism. Island sustainability indicators may be categorised based on economic, environmental, and socio-cultural indicators (Polnyotee & Thadaniti, 2014). Economic indicators include accommodation, human resources, finance, inter-sectoral linkages, and transportation (United Nations World Tourism Organization, 2004). Environmental indicators include energy and water consumption, waste disposal, environmental planning, and projects (Grilli,

Tyllianakis, Luisetti, Ferrini, & Turner, 2021). Socio-cultural indicators relate to cultural stress, social ills, behavioural characteristics of hosts and tourists (Agyeiwaah, McKercher, & Suntikul, 2017). Island sustainability indicators have been viewed as either having a positive or a negative impact. Understanding the role of sustainability indicators in sustainable island tourism development requires elaboration.

1.4 Sustainable island development

Sustainable development suggests that islands will be in existence on an ongoing basis and that citizens and residents are afforded an opportunity to work and live reasonably well on islands. The Sustainable Development Goals (SDGs) have been developed to set in place specific achievement targets that would result in sustainable development and the 17 SGDs are applicable to the achievement of sustainable island tourism (Fauzel & Tandrayen-Ragoobur, 2023). Islands are primarily dependent on tourism as a sector for economic growth (Wilkinson, 1987). Sustainable development on an island must be unpacked to realize a vision of sustainability. The future of island tourism sustainability is increasingly difficult, as islands face myriads of challenges that take sustainable development in several directions (Parra-López, Barrientos-Báez, & Sánchez, 2023). A series of natural and economic disasters may place vulnerable islands on a downward path that reverses any advancement towards sustainable development (McLeod, 2022).

Island governments are caught between whether to continue a path of tourism development or diversify. Island tourism resilience has been seen as an inbuilt capacity to rebound in a tourism sector that may have sustained continued growth (McLeod, 2020). Building certain structures, systems, and strategies that will allow resilience to external shocks and controlling any downward spirals are steps for further growth of the tourism sector. Sustainable development also suggests that an island may achieve developed status. Being an area that is relatively small with smaller population sizes than metropolitan areas, presumably, the management of resources should be easily achieved. Nevertheless, island nations have been subsumed with growing debt and poverty that suggests that growth built on tourism may be questionable (Higgins-Desbiolles, 2022). Understanding the contribution of tourism to sustainable development must be unpacked.

The impacts of island tourism suggest that three categories of policy and management issues should be addressed. As such, this book is structured into three parts (Figure 1.2). Part I details *Island Tourism Policy* as four chapters about demand, supply, sustainability, and policy processes. Part II contains *Island Tourism Governance* and includes content about governance structures, systems, strategies, and coordination and consensus-building. Part III outlines *Island Tourism Planning for Sustainable Development* and considers planning frameworks, resources, sustainable development, and sustainable development futures. The book concludes with charting a way forward for island tourism policy, governance, and sustainable development.

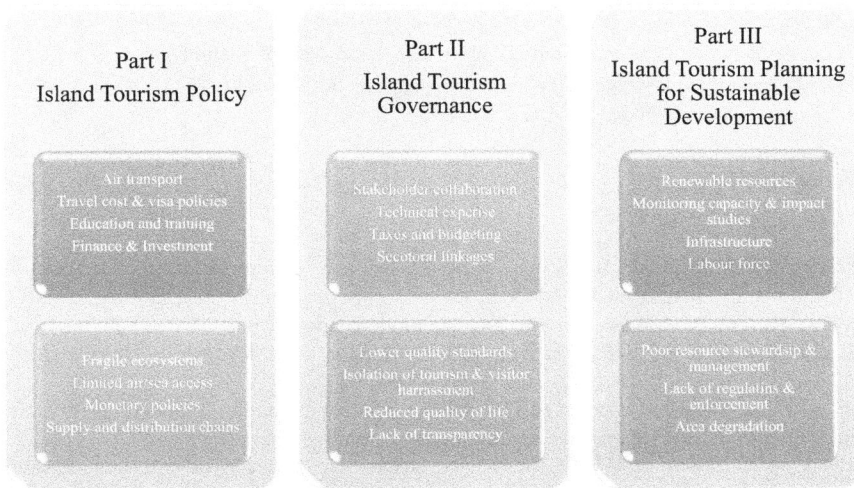

Figure 1.2 A framework of policy and management issues in island tourism.

1.5 Conclusion

The sustainable development of islands must be undertaken through careful analysis of the impacts of tourism development on islands. The relationship between islands and tourism is never ending as islands are popular tourist destinations (Harrison, 2001). Island characteristics indicate that tourism is built on fragile ecosystems that must be regenerated and nurtured for sustainability (Gounder, 2022). Considering the physical and human characteristics of islands outlined in this chapter, the tourism sector must be integrated into an island's economy in a beneficial way, wherein the benefits of tourism must outweigh the costs. To understand steps towards the sustainable development of island tourism, this book takes a direction that considers tourism policy, governance, planning for sustainable development.

Chapter 1 discussion questions

1 Compare and contrast the characteristics of islands in warm and cold weather destinations.
2 The larger the island, the greater the tourism impacts. Consider this statement with a comparative analysis of tourism's impacts on two islands of different land areas and population sizes.

3 Seasonality affects the scale of positive and negative impacts of tourism on islands. Discuss this statement with case examples of islands showing both positive and negative impacts from tourism activities in the peak and off seasons.

References

Agyeiwaah, E., McKercher, B., & Suntikul, W. (2017). Identifying core indicators of sustainable tourism: A path forward? *Tourism Management Perspectives, 24*(October), 26–33.

Buckley, R. (2002). Surf tourism and sustainable development in Indo-Pacific Islands. II. Recreational capacity management and case study. *Journal of Sustainable Tourism, 10*(5), 425–442.

Butler, R. W. (1980). The concept of a tourist area cycle of evolution: Implications for management of resources. *Canadian Geographer/Le Géographe canadien, 24*(1), 5–12.

Carlsen, J., & Butler, R. (2011). Introducing sustainable perspectives of island tourism. In J. Carlsen & R. Butler (Eds.), *Island tourism: Sustainable perspectives* (pp. 1–7). Wallingford, UK: CABI.

Cohen, C. B. (1995). Marketing paradise, making nation. *Annals of Tourism Research, 22*(2), 404–421.

Croes, R. (2011). *The small island paradox: Tourism specialization as a potential solution.* London: Lambert Academic Publishing.

Croes, R. (2006). A paradigm shift to a new strategy for small island economies: Embracing demand side economics for value enhancement and long term economic stability. *Tourism Management, 27*(3), 453–465.

Eriksen, T. H. (2020). Implications of runaway globalisation in the Seychelles. *Small States & Territories, 3*(1), 9–20.

Eslami, S., Khalifah, Z., Mardani, A., & Streimikiene, D. (2018). Impact of non-economic factors on residents' support for sustainable tourism development in Langkawi Island, Malaysia. *Economics & Sociology, 11*(4), 181–197.

Fauzel, S., & Tandrayen-Ragoobur, V. (2023). Sustainable development and tourism growth in an island economy: A dynamic investigation. *Journal of Policy Research in Tourism, Leisure and Events, 15*(4), 502–512.

Gounder, R. (2022). Tourism-led and economic-driven nexus in Mauritius: Spillovers and inclusive development policies in the case of an African nation. *Tourism Economics, 28*(4), 1040–1058.

Grilli, G., Tyllianakis, E., Luisetti, T., Ferrini, S., & Turner, R. K. (2021). Prospective tourist preferences for sustainable tourism development in Small Island Developing States. *Tourism Management, 82*(February), 104178.

Hampton, M. P., & Christensen, J. (2007). Competing industries in islands a new tourism approach. *Annals of Tourism Research, 34*(4), 998–1020.

Harrison, D. (2001). Islands, image and tourism. *Tourism Recreation Research, 26*(3), 9–14.

Higgins-Desbiolles, F. (2022). The ongoingness of imperialism: The problem of tourism dependency and the promise of radical equality. *Annals of Tourism Research, 94*(May), 103382.

Kothari, U., & Arnall, A. (2020). Shifting sands: The rhythms and temporalities of island sandscapes. *Geoforum, 108*(January), 305–314.

Mason, P. (2020). *Tourism impacts, planning and management*. Oxon: Routledge.

McLeod, M. (2020). Tourism governance, panarchy and resilience in the Bahamas. In S. Rolle, J. Minnis. & I. Bethell-Bennett (Eds.), *Tourism development, governance and sustainability in the Bahamas* (pp. 103–113). Abingdon, UK: Routledge.

McLeod, M. (2022). Tourism Destination Recovery, a Case Study of Grand Bahama Island. In I. Bethell-Bennett, S. Rolle, J. Minnis, & F. Okumus (Eds.), *Pandemics, Disasters, Sustainability, Tourism* (pp. 93–108). Leeds, UK: Emerald.

McLeod, M., & Croes, R. R. (2018). *Tourism management in warm-water island destinations* (Vol. 6). Wallingford, UK: CABI.

McLeod, M., Dodds, R., & Butler, R. (2022). *Island tourism sustainability and resiliency*. Abingdon, UK: Routledge.

Mintz, S. (1965). The Caribbean as a socio-cultural area. *Cahiers d'Histoire Mondiale. Journal of World History. Cuadernos de Historia Mundial, 9*(1), 912.

Moyle, B., Croy, G., & Weiler, B. (2010). Tourism interaction on islands: The community and visitor social exchange. *International Journal of Culture, Tourism and Hospitality Research, 4*(2), 96–107.

National Geographic (2023). Island. Retrieved from https://education.nationalgeographic.org/resource/island/

Oroz, T. (2022). Multiple island temporalities: "Island time" and the spatialisation of slowness on the Dalmatian Island of Dugi otok. *Narodna umjetnost: hrvatski časopis za etnologiju i folkloristiku, 59*(2), 9–38.

Parra-López, E., Barrientos-Báez, A., & Sánchez, M. d. l. Á. P. (2023). Island destinations in the face of global challenges. In A. Morrison & D. Buhalis (Eds.), *Routledge handbook of trends and issues in tourism sustainability, planning and development, management, and technology* (pp. 151–159). Abingdon, UK: Routledge.

Petersen, G. (2009). *Traditional micronesian societies: Adaptation, integration, and political organization in the central Pacific*. Hawai'i: University of Hawai'i Press.

Polnyotee, M., & Thadaniti, S. (2014). The survey of factors influencing sustainable tourism at Patong beach, Phuket island, Thailand. *Mediterranean Journal of Social Sciences, 5*(9), 650–650.

Sánchez-Cañizares, S. M., & Castillo-Canalejo, A. M. (2014). Community-based island tourism: The case of Boa Vista in Cape Verde. *International Journal of Culture, Tourism and Hospitality Research, 8*(2), 219–233.

Sealey, K. S., McDonough, V. N., & Lunz, K. S. (2014). Coastal impact ranking of small islands for conservation, restoration and tourism development: A case study of The Bahamas. *Ocean & Coastal Management, 91*(April), 88–101.

Sheldon, P. J. (2005, October, 2005). *The challenges to sustainability in island tourism*. Occasional Paper, (1, 2005-01). Hawai'i.

Stratford, E. (2016). *Island geographies: Essays and conversations*. Abingdon, UK: Routledge.

United Nations World Tourism Organization (2004). *Indicators of sustainable development for tourism destinations*. Madrid: World Tourism Organization.

Wilkinson, P. F. (1987). Tourism in small island nations: A fragile dependence. *Leisure Studies, 6*(2), 127–146.

World Bank (2023). World development indicators data bank. Retrieved from https://databank.worldbank.org/source/world-development-indicators.

Part I
Island tourism policy

2 Tourism policies to influence demand

2.1 Introduction

This chapter outlines a framework to guide the development of tourism policies to influence demand for island destinations. Policies relating to managing tourism demand for island destinations are wide and varied. In most instances, policies to drive tourism demand are dispersed over government institutions with no responsibility for tourism development. In such circumstances, those public policies affecting tourism development need to be understood. Early writings about tourism demand in islands, during the 1970s, sought to understand the impact and potential of tourism (Boissevain, 1979; LaFlamme, 1979; Peters, 1980). Generally, determinants of tourism demand have been explored as relating to exogenous, social-psychological and economic factors (Uysal, 1998). Croes and Ridderstaat (2018) have considered tourism motivation and demand for islands and have argued that a congruence between the tastes and preferences of the country of origin and the destination is necessary to drive tourism demand.

Unpacking tourism demand for islands across the globe with different geological, geographical, historical, and economic contexts requires an understanding of both outbound flows from countries of origin and inbound flows to destinations. Tourism demand to island destinations has a myriad of challenges. From the perspective of outbound flows, emerging drivers of tourism demand towards island destinations include climate, business cycles, money supply, quality of life, diseases, and vegetation issues (Croes & Ridderstaat, 2018). From the perspective of inbound flows, seasonality poses a major tourism demand challenge for island destinations (Liasidou, Garanti, & Pipyros, 2022). Andriotis (2005) has recommended product mix diversification, customer mix changes, and aggressive pricing to overcome a seasonality problem in Crete. While tourism demand may be created, getting to remote islands is another challenge. The television series *Fantasy Island* from 1977 to 1984 is reminiscent of the importance of air transport 'the plane' to islands (Johnston, 2022). Air seat capacity is a prime resource for island destinations (Liasidou et al., 2022) and low-cost airlines have been an essential part of the airlift mix (Álvarez-Díaz, González-Gómez, & Otero-Giráldez, 2019).

Chapter 2 is divided into three sections to explore the elements of tourism demand for island destinations. The first section details the tourist motivation to visit

DOI: 10.4324/9781003435112-3

islands and proposes a hierarchical framework of tourism demand characteristics. The second section considers the influences on island tourism demand to understand how tourism demand is maintained for island destinations that are primarily dependent on tourism. The third section outlines policies of tourism demand including air transport, product development, and general macro-economic policies. Facilitating conditions for travel to islands including, visa policies, travel cost, airlift capacity, and knowledge about island destinations are considered. A conclusion follows this, with the key elements for the achievement of sustainable tourism demand for island destinations.

2.2 Tourist motivation for island destinations

Islands are popular tourist destinations, as exotic flora and fauna and a relaxed way of life appeal to visitors from across the globe. Escapism, nostalgia, and reconnection are push factors of tourism demand (Dann, 2012), and the lure of the lush scenery brings about a pull to island destinations (Litvin & Ling, 2001). Understanding the motives of visitors to island destinations is complicated by several intervening factors. The social dynamics of a visit may contribute more to visiting an island rather than an actual intrinsic motive of leisure or relaxation, and then there are spontaneous actions that may not have been related to a planned visit (Larsen, Urry, & Axhausen, 2007). Larsen et al. (2007) have argued that tourist motivation includes social motives involving meeting family obligations and maintaining friendship bonds. In the context of island tourism, social motives of travel need to be fully explored to understand the range of activities that are based on socialization. With limited economic opportunities, island populations migrate thereby creating opportunities for a thriving visiting friend and relatives' market (VFR) (Gibson, Pratt, & Iaquinto, 2022). The VFR market has remained largely untapped and requires more research into the motives for travel.

Island tourism demand can be understood from various perspectives including an understanding of island identity and the pull to island environments. The imagery of islands is alluring (Harrison, 2001), but imagery alone may not develop that desire to visit an island destination. The economic, socio-psychological, and the general business environment's support for visiting a destination are important considerations (Page, 2014). As such, characteristics of tourism demand are viewed from the elements of a tourist experience. A tourist experience can be related to key questions within a hierarchical framework of tourism demand characteristics (Figure 2.1). The first set of characteristics relates to who the tourist is (tourist typology in terms of demographic characteristics), why the tourist travels (motivation and purpose of visit), and when the tourist travels (time, money, push and pull factors). The second set of characteristics relates to how (air, sea, land transport) and where the tourist travels (long or short haul), a relationship of distance and accessibility (geographic characteristics). The third set of characteristics relates to what the tourist does at the destination in terms of products, experiences, and events (Figure 2.1).

Understanding who the tourist is clarifies the nature of tourism demand. Cohen (1974) has suggested that there are different types of tourists based on certain dimensions of direction, distance, general purpose and specific purpose,

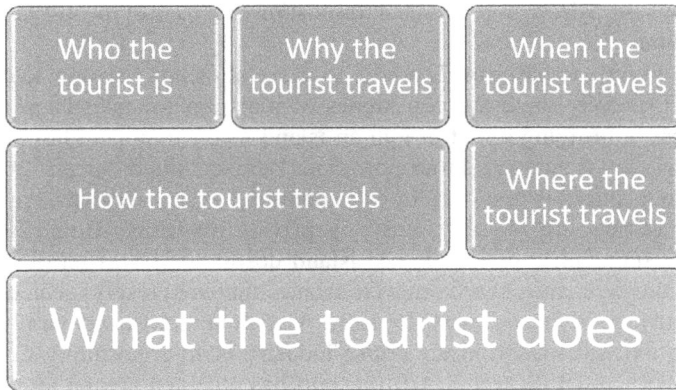

Figure 2.1 A hierarchical framework of tourism demand characteristics.

permanency, recurrency, and voluntariness, and has argued that the primary motive for tourism activities relates to 'novelty and change'. Island destinations are seen as exotic landscapes (Minca, 2000), with scenic beaches and culturally diverse indigenous populations. Whether island landscapes can sustain the need for novelty and change is debatable. Although specific types of tourists, the wanderer and explorer, have been associated with island tourism (Buckley, 2002), economic sustainability is built on the masses of visitors to island destinations. Plog (2001) and Cohen (1974) have detailed types of tourists, however, an island tourist typology requires elaboration. A particular tourist's disposition that results in the selection and visitation to an island destination is important for sustainability. The dismissive notion that islands are fragile and vulnerable (Farbotko, 2010), and therefore sustainability is challenged when visitors move to other areas may not be a reality. Rather, the creation of an island environment, even man-made, will attract visitors based on the idea of the 'exotic' inherently an attribute of an island.

The reason as to why a tourist travels is complex (Dellaert, Arentze, & Horeni, 2014). Tourist travel motivation theories include benefits sought or realized approaches, expectancy-based approaches, needs-based approaches, push-and-pull approaches, and values-based approaches (Page, 2014). Push-and-pull motivation simplifies understanding about tourist motivation to travel to islands. Carvache-Franco, Carvache-Franco, and Hernández-Lara (2021) have analysed tourist motivational factors to visit the Galápagos Islands in Ecuador and have found that the beach experience, culture, authenticity, and nature are key motivational factors. The reason as to why may be interrelated to when a tourist travels, as the reason can be time bound. In addition, a travel time dimension has a cost factor and it depends on a tourist's budget, and if a tourist has limited time, a destination that is closer may be selected as suggested by Fennell (1996). Visiting an island may largely depend on the cost of the visit as tourism's actual demand requires time and money (Yun, Hao, Si, & Mei, 2023). Time also relates to the time of year, although it is often viewed as a matter of seasonality, time of year is also related to cost and length of stay factors. Almeida, Machado, and Xu (2021) have found that the length of stay has been declining for island destinations because of the availability of low-cost airlines.

The matter of how a tourist travels relates to the relationship between tourism and transport. Primarily there are two modes of transport for islands: air and sea. Nonetheless, a bridging effect to join an island to the mainland has possibilities for improving island travel transportation (Baldacchino & Spears, 2007). Sea transport to islands has received limited attention as ferry services, cruises, and other sea vessels are mostly owned by private investors (Bola, 2017). Although the sustainability of cruising to island destinations was particularly evident during the pandemic, sea voyages to islands that are closely located and involve a day trip have the potential to grow. Ajagunna and Casanova (2022) have suggested that shifting from the cruise industry is an opportunity for the Caribbean's luxury yachting sector. Air transport has received greater attention in island tourist destinations. Mazzola, Cirà, Ruggieri, and Butler (2022) have suggested examination of air transport costs to understand the motivation of tourists to visit a specific island destination. In a critical review of the tourism–transport relationship in Madeira, Barros (2012) has suggested a relationship between type of tourist and type of air transport. Schedule air travellers were determined based on age, beach, culture, and travel agency website; low-cost travellers were determined based on gender, expenditure, airline, or travel agency website; and charter-flight travellers were determined based on age, expenditure, and culture (Barros, 2012).

Destination selection relates to where to travel. Parra-López and Martínez-González (2018) in a review of island destinations noted that tourists' selection of islands is based on island characteristics. Where a tourist travels to and what a tourist does go hand in hand as a chosen island has features that will provide the basis for an experience. Yiamjanya and Wongleedee (2014) have noted natural scenery including beaches and the weather as an important motivator to visit islands. Liang (2017) has concurred and has noted that islands are promoted in different ways, however, beach and water- related activities are the dominant experiences. As such, warm weather island destinations have benefited from these motivations. Almeida and Garrod (2018) have gone further and have proposed that for mature island destinations both familiarity and novelty are important to chart a way forward for product reinvention.

2.3 Influences on island tourism demand

An influence on tourism demand is an action that increases or decreases demand. Strategic tourism marketing activities seek to influence tourism demand in various ways. Muryani, Permatasari, and Padilla (2020) have shown in the case of Indonesia, increasing tourism promotion funds is a strategy readily adopted to increase visitor arrivals. Branding of island destinations is an important consideration to distinguish one island from another. An island destination that develops a unique selling proposition and brand successfully influences tourism demand

(Cave & Brown, 2012). On one hand, Liang (2017) has found that a brand advantage among island destinations was not as important as the similarity of tourism activities on island in promoting island tourism, and has argued for co-branding of island destinations. On the other hand, the differences in the physical features and activities of islands make one destination brand proposition challenging. Visitor expectations and realization of their experiences of visiting islands are based on marketing messages. Demographic characteristics of visitors to island destinations are important for understanding how to reach visitors that fit the profile with marketing messages (Park, Hsieh, & McNally, 2010). As such, message and media strategies for marketing island destinations should consider the indigenous legacies of islands to offer authenticity and inclusivity in marketing practices (Walker, 2020). Indigenous marketing practices are essential to sustain the culture and heritage of islands, and the way of life provides rich experiences for visitors (Cameron & Gatewood, 2008).

Economic factors have a direct influence on island tourism demand. Income is an important consideration of tourism demand for island destinations. Takahashi (2020) has argued that an island that is dependent on tourism with a less diversified economy such as the French Polynesia attracts a high-income market in comparison with Singapore with economic diversity that attracts several markets. Income elasticity, price, and other economic determinants influence international demand for island tourism (Inchausti-Sintes, Voltes-Dorta, & Suau-Sánchez, 2021; Kumar, Kumar, Patel, Hussain Shahzad, & Stauvermann, 2020). In addition, considering the effect of global economic uncertainty has implications for island destinations. Nguyen, Schinckus, and Su (2020) have pointed out that outbound tourism from tourist-generating markets is directly related to the global macroeconomic policy uncertainty. Understanding uncertainty and its effect on visitor numbers is important for policies that will influence demand from particular tourist-generating markets to remote islands (Shareef & McAleer, 2008).

With consideration that putting all the eggs in the tourism basket is risky for island destinations, tourism specialization has its advantages. Croes (2022, p. 39) has argued for tourism specialization in islands and has pointed out that such a policy has two advantages: 'trade transaction costs are lowered with increased trust' and 'reduced production costs' that are relatively high in small islands. Arguably, measures can be taken to manage any risk of tourism specialization as indicators and predictors for changes in island tourism demand can avert crises that may result in economic failure. Rosselló-Nadal (2001) has forecasted turning points in the cyclical evolution of tourism demand from the UK and Germany to the Balearic Islands. Soh, Puah, and Arip (2019) have tested a tourism cycle indicator (TCI) in the Maldives and determined a lead time of 4.4 months with 10 turning point events, including economic crisis and political instability. Such a lead time prediction can assist island governments with taking actions to avert troughs in tourism demand (Soh et al., 2019). A tourism demand cycle approach, which forecasts changes in demand of the main markets to island destinations, assists with making decisions about market selection and marketing resource allocation. A case study about tourism in Hawai'i follows to explore tourism demand issues in islands.

Mini Case Study 1: Hawaiians want tourists to leave

In 2021, the Annual Visitor Research Report for Hawai'i indicates that 6,777,760 airline visitors came to the islands with a total visitor expenditure of US$ 13.15 billion, and average length of stay of 9.64 days (Hawai'i Tourism Authority, 2021). Most visitors arrived from the US West Coast and the next largest number of visitors was from the US East Coast (Hawai'i Tourism Authority, 2021). The main purpose of visit was vacation, followed by visiting friends and relatives and honeymoon, respectively (Hawai'i Tourism Authority, 2021). The 2021 visitor satisfaction survey indicated that satisfaction was highest for US West and East coast travellers under the age of 35 years, and less affluent travellers for their most recent trip (State of Hawai'i, 2021). Visitors from the West Coast were most likely to return to Hawai'i (81.6%) than visitors from the East Coast (65.9%), however, the top reasons for not returning were too expensive (34.2%), want to go someplace new (32.5%), and poor value (26.0%) (State of Hawai'i, 2021).

At the rebound from the pandemic, several online reports circulated that Hawai'i's residents were wary of tourism (Dethlefsen, 2023; McDonagh, 2022; Rodriguez, 2023). Native Hawaiians have long wanted tourism numbers to decrease to the islands and this is evident by the reduced number of natives working in the tourism sector (Rodriguez, 2023). Native Hawaiians comprised the majority of the board of the Hawai'i Tourism Authority and have made a change in focus from destination marketing to destination management (Dethlefsen, 2023). According to Dethlefsen (2023), Honolulu is the seventh most-visited city in the United States, and in 2019, 216,000 jobs directly depended on tourism. Although heavily dependent on tourism, residents have negative sentiments about tourism because of water shortages, tourism energy consumption, and contributing affordable housing crisis (Dethlefsen, 2023; McDonagh, 2022). With tourists renting U-hauls and camping on the beach (Dethlefsen, 2023), and shortages of hospitality workers (McDonagh, 2022), tourism demand in Hawai'i far outstrips supply.

Mini Case Study 1 Discussion Questions

1 What are the influences driving tourism demand in Hawai'i?
2 Why should Hawai'i tourism policymakers be concerned about a high tourism demand?
3 What are the policies that will reduce tourism demand in Hawai'i?

2.4 Policies for island tourism demand

Islands are surrounded by water and can only be reached through air or sea transportation. Policies about island travel and transportation have received limited attention (Mazzola et al., 2022). While air transportation liberalization and deregulation policies for island destinations have been considered by Papatheodorou (2000) to some extent, evidence is needed to further support a justification for broad brush liberalization and deregulation of island air space. An argument in support of Open Skies Agreements to liberalize air transport is based on a reduction in transport costs (Micco & Serebrisky, 2006). One of the earliest Open Skies Agreements was signed between the Netherlands and the United States of America in 1992 to facilitate airlift between the USA and Aruba (de Leon, 2002). In an overview of island destinations, Piermartini and Rousová (2013) found that liberalization of air services has a positive effect on passenger flows. An Air Liberalization Index (ALI) showed the rank, in brackets, of islands as follows: Sao Tomo and Principe (7), Bahamas (19), Solomon Islands (20, and Seychelles (25), Mauritius (31), Samoa (41), Cuba (47), and Fiji (56) (Piermartini & Rousová, 2013). Islands are isolated, and tourism is dependent on the air seat capacity. With supportive air services, a tourism industry will thrive. Within an Open Skies Policy, air service liberalization should also consider the mix of scheduled, cargo, and charter aircrafts serving the destination (Cordes, 1993). Álvarez-Díaz et al. (2019) have considered whether low-cost carriers (LCCs) increase tourism demand. Sensitivity of tourism demand to price changes suggests that LCCs may contribute to tourism demand expansion (Álvarez-Díaz et al., 2019).

Airlift capacity is critical for island tourism. The air capacity issue for islands is one of economies of scale. For example, US airlines may want greater access to a larger market for operations (Cordes, 1993). In addition, environmental considerations of air pollution and carbon emissions must be considered within the context of an Open Skies Policy. Duval (2013) has considered the relationship between transport and tourism and has suggested the adoption of climate policies and regulatory restrictions with an acknowledgment of the effect for tourist flows. Nonetheless, in a small island context, multi-lateral agreements to build airlift capacity has a definite advantage. Yarde and Jönsson (2016) have found that a regional multi-lateral agreement will improve intra-regional tourism. In an intra-regional context, airline survival is dependent on the profitability of routes, and with competition from international carriers regional governments have to heavily subsidize regionally based carriers (Yarde & Jönsson, 2016). Subsidization of air transport is unsustainable and as noted by Kumar and Patel (2023) international air travel declines can hamper economic growth in the island of Fiji. Liasidou (2017) has suggested that tourism policy should be drafted based on strategic synergies and links with airlines. Air transport policy is critical for tourism sustainability in island destinations, and national airlines form part of such a policy.

An island tourism demand policy framework is proposed (Figure 2.2). Based on a tourism demand analysis of Hawai'i, Yun et al. (2023) have recommended that particular attention be given to the pricing of tourism products and services, increasing airline seat capacity, improving service quality, optimizing staff recruitment and training, and promoting the destination effectively. Lim and Zhu (2017) have shown evidence in the case of Singapore that integrated resorts (IRs) contribute to tourism demand growth in attracting tourists from Asia. Integrated resorts are developed clusters of tourism products and services including accommodation,

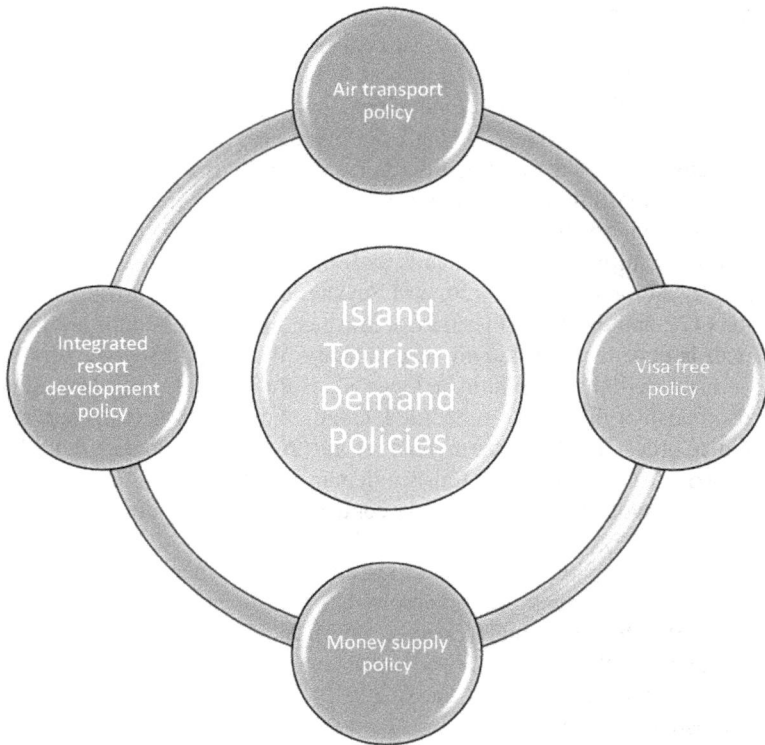

Figure 2.2 Island tourism demand policy framework.

attractions, and other amenities. Policies to encourage resort style development as opposed to other types of tourism development are important to attract tourists to island destinations. Policies of market diversification and a special visa waiver programme for non-US visitors are recommended in the case of Puerto Rico that receives 90% of its visitors from the US market (Husein & Kara, 2020). A free entry visa policy implemented in Indonesia in 2003 for Asian countries had a positive effect on tourism demand (Muryani et al., 2020).

Finally, economic policies are primarily a driver of tourism growth and development. Kim, Lee, and Mjelde (2018) have argued that Abenomics increased tourist arrivals from South Korea to Japan. Abenomics is an expansionary monetary policy of increasing the money supply, increasing government spending and reforms (Fukuda, 2015). An expansionary monetary policy must be considered in context. Inflation and price increases of tourism products and services in a small island may result from an expansionary monetary policy. Monetary policy to curb inflation and visa-free travel policies are important to facilitate air travel to islands (Kumar & Patel, 2023). In addition, the effects of exchange rate policy on tourism demand require greater attention. Chi (2020) has found that outbound travellers from South Korea select destinations with stable exchange rates.

2.5 Conclusion

This chapter explores island tourism demand and underscores a supportive policy framework to sustain tourism demand. Agbola, Dogru, and Gunter (2020) have argued that understanding the drivers of tourism demand is critical in a post-pandemic age. Understanding tourists' travel behaviour, their preferences, and selection of islands guide policies relating to tourism demand. Demand for island tourism products and services relates to understanding the tastes and preferences for traveling to island destinations (Croes & Ridderstaat, 2018). The literature suggests that an effective air transport policy supports tourism demand for islands (Kumar & Patel, 2023). Sustainability of tourism demand is related to air transport agreements that islands have signed unto (Yarde & Jönsson, 2016). Changes in tourism demand must be monitored by using indicators and forecasts (Rosselló-Nadal, 2001; Soh et al., 2019). An overview of this chapter shows the important drivers of tourism demand for islands including income, time, and destination characteristics, facilitated by airlift, resort development, monetary and entry policies, and these elements are critical for island tourism sustainability. Based on a push-and-pull concept of tourism motivation, the tourism product element matched with the characteristics of a tourist is important to enhance tourist perception of islands and motivate a visit. In addition, information and disinformation must be monitored to ensure tourist expectations are being effectively managed. An island tourism demand policy framework is proposed. Finally, in times of crisis such as natural disasters or diseases, temporal changes in tourism demand for islands are to be planned for and managed.

Chapter 2 discussion questions

1 Using the hierarchical framework of tourism demand characteristics, populate each block with specific characteristics for a small island and specific demand characteristics for a large island.
2 Why do tourists travel to islands and what will reduce the number of tourists visiting islands?
3 Discuss the island tourism demand policy framework using an example of an island that has received a decline in tourist arrivals over the last 5 years.

References

Agbola, F. W., Dogru, T., & Gunter, U. (2020). *Tourism demand: Emerging theoretical and empirical issues* (Vol. 26, pp. 1307–1310). London: Sage Publications.

Ajagunna, I., & Casanova, S. (2022). An analysis of the post-COVID-19 cruise industry: Could this be a new possibility for the luxury yacht sector in the Caribbean? *Worldwide Hospitality and Tourism Themes, 14*(2), 115–123.

Almeida, A., & Garrod, B. (2018). A CATREG model of destination choice for a mature Island destination. *Journal of Destination Marketing & Management, 8*(2018), 32–40.

Almeida, A., Machado, L. P., & Xu, C. (2021). Factors explaining length of stay: Lessons to be learnt from Madeira Island. *Annals of Tourism Research Empirical Insights, 2*(1), 100014.

Álvarez-Díaz, M., González-Gómez, M., & Otero-Giráldez, M. S. (2019). Low cost airlines and international tourism demand. The case of Porto's airport in the northwest of the Iberian Peninsula. *Journal of Air Transport Management, 79*(August), 101689.

Andriotis, K. (2005). Seasonality in Crete: Problem or a way of life? *Tourism Economics, 11*(2), 207–224.

Baldacchino, G., & Spears, A. (2007). The bridge effect: A tentative score sheet for Prince Edward Island. In G. Baldacchino (Ed.), *Bridging islands: The impact of "fixed links"* (pp. 49–68). Charlottetown: Acorn Press.

Barros, V. G. (2012). Transportation choice and tourists' behaviour. *Tourism Economics, 18*(3), 519–531.

Boissevain, J. (1979). The impact of tourism on a dependent island: Gozo, Malta. *Annals of Tourism Research, 6*(1), 76–90.

Bola, A. (2017). Potential for sustainable sea transport: A case study of the Southern Lomaiviti, Fiji islands. *Marine Policy, 75*(January), 260–270.

Buckley, R. (2002). Surf tourism and sustainable development in Indo-Pacific Islands. II. Recreational capacity management and case study. *Journal of Sustainable Tourism, 10*(5), 425–442.

Cameron, C. M., & Gatewood, J. B. (2008). Beyond sun, sand and sea: The emergent tourism programme in the Turks and Caicos Islands. *Journal of Heritage Tourism, 3*(1), 55–73.

Carvache-Franco, W., Carvache-Franco, M., & Hernández-Lara, A. B. (2021). From motivation to segmentation in coastal and marine destinations: A study from the Galapagos Islands, Ecuador. *Current Issues in Tourism, 24*(16), 2325–2341.

Cave, J., & Brown, K. G. (2012). Island tourism: Destinations: An editorial introduction to the special issue. *International Journal of Culture, Tourism and Hospitality Research, 6*(2), 95–113.

Chi, J. (2020). The impact of third-country exchange rate risk on international air travel flows: The case of Korean outbound tourism demand. *Transport Policy, 89*(April), 66–78.

Cohen, E. (1974). Who is a tourist?: A conceptual clarification. *The Sociological Review, 22*(4), 527–555.

Cordes, J. H. (1993). Flying the open skies: An analysis and historical perspective of the US–Netherlands bilateral air transport agreement of September 4, 1992. *Transnat'l Law, 6*(1), 301–327.

Croes, R. (2022). *Small island and small destination tourism: Overcoming the smallness barrier for economic growth and tourism competitiveness.* Abingdon, UK: CRC Press.

Croes, R., & Ridderstaat, J. (2018). Tourist motivation and demand for islands. In *Tourism management in warm-water island destinations: Systems and strategies* (pp. 44–62). Wallingford, UK: CABI.

Dann, G. M. (2012). Nostalgia in the noughties. In W. F. Theobald (Ed.), *Global tourism* (pp. 32–51). Burlington, UK: Routledge.

de Leon, P. M. (2002). Before and after the tenth anniversary of the Open Skies Agreement Netherlands–US of 1992. *Air and Space Law, 27*(4/5). Retrieved from www.washing tonpost.com/archive/business/1992/09/05/us-netherlands-agree-to-open-skies/2e2b26d9-c156-4ebe-9517-ea327f654062/

Dellaert, B. G., Arentze, T. A., & Horeni, O. (2014). Tourists' mental representations of complex travel decision problems. *Journal of Travel Research, 53*(1), 3–11.

Dethlefsen, M. (2023, February 15, 2023). The case for caps: Overtourism in Hawaii. Retrieved from https://brownpoliticalreview.org/2023/02/tourism-in-hawaii-the-case-for-caps/

Duval, D. T. (2013). Critical issues in air transport and tourism. *Tourism Geographies, 15*(3), 494–510.

Farbotko, C. (2010). Wishful sinking: Disappearing islands, climate refugees and cosmopolitan experimentation. *Asia Pacific Viewpoint, 51*(1), 47–60.

Fennell, D. A. (1996). A tourist space-time budget in the Shetland Islands. *Annals of Tourism Research, 23*(4), 811–829.

Fukuda, S.-i. (2015). Abenomics: Why was it so successful in changing market expectations? *Journal of the Japanese and International Economies, 37*(September), 1–20.

Gibson, D., Pratt, S., & Iaquinto, B. L. (2022). Samoan perceptions of travel and tourism mobilities – The concept of Malaga. *Tourism Geographies, 24*(4–5), 737–758.

Harrison, D. (2001). Islands, image and tourism. *Tourism Recreation Research, 26*(3), 9–14.

Hawai'i Tourism Authority (2021). *2021 Annual Visitor Research Report.* Hawaii: Hawaii Tourism Authority. Retrieved from www.hawaiitourismauthority.org/media/9691/2021-annual-report-revfinal-with-cover.pdf

Husein, J., & Kara, S. M. (2020). Nonlinear ARDL estimation of tourism demand for Puerto Rico from the USA. *Tourism Management, 77*(April), 103998.

Inchausti-Sintes, F., Voltes-Dorta, A., & Suau-Sánchez, P. (2021). The income elasticity gap and its implications for economic growth and tourism development: The Balearic vs the Canary Islands. *Current Issues in Tourism, 24*(1), 98–116.

Johnston, D. (2022). "The plane! The plane!": 45 years ago, this underrated TV show perfected the anthology format. Retrieved from www.inverse.com/entertainment/fantasy-island-45th-anniversary.

Kim, J., Lee, C., & Mjelde, J. (2018). Impact of economic policy on international tourism demand: The case of Abenomics. *Current Issues in Tourism, 21*(16), 1912–1929.

Kumar, N., Kumar, R. R., Patel, A., Hussain Shahzad, S. J., & Stauvermann, P. J. (2020). Modelling inbound international tourism demand in small Pacific Island countries. *Applied Economics, 52*(10), 1031–1047.

Kumar, N. N., & Patel, A. (2023). Nonlinear effect of air travel tourism demand on economic growth in Fiji. *Journal of Air Transport Management, 109*(June), 102402.

LaFlamme, A. G. (1979). The impact of tourism: A case from the Bahama Islands. *Annals of Tourism Research, 6*(2), 137–148.

Larsen, J., Urry, J., & Axhausen, K. W. (2007). Networks and tourism: Mobile social life. *Annals of Tourism Research, 34*(1), 244–262.

Liang, A. R.-D. (2017). Assessing the impact of co-branding of island destination and tourism activities on tourists' reactions. *Current Issues in Tourism, 20*(5), 536–551.

Liasidou, S. (2017). Drafting a realistic tourism policy: The airlines' strategic influence. *Tourism Review, 72*(1), 28–44.

Liasidou, S., Garanti, Z., & Pipyros, K. (2022). Air transportation and tourism interactions and actions for competitive destinations: The case of Cyprus. *Worldwide Hospitality and Tourism Themes, 14*(5), 470–480.

Lim, C., & Zhu, L. (2017). Dynamic heterogeneous panel data analysis of tourism demand for Singapore. *Journal of Travel & Tourism Marketing, 34*(9), 1224–1234.

Litvin, S. W., & Ling, S. N. S. (2001). The destination attribute management model: An empirical application to Bintan, Indonesia. *Tourism Management, 22*(5), 481–492.

Mazzola, F., Cirà, A., Ruggieri, G., & Butler, R. (2022). Air transport and tourism flows to islands: A panel analysis for southern European countries. *International Journal of Tourism Research, 24*(5), 639–652.

McDonagh, S. (2022). Hawaii overtourism: Residents beg tourists to stop visiting amid post-pandemic boom. Retrieved from www.euronews.com/travel/2022/05/03/hawaiian-overtourism-residents-beg-tourists-to-stop-visiting-amid-post-pandemic-boom

Micco, A., & Serebrisky, T. (2006). Competition regimes and air transport costs: The effects of open skies agreements. *Journal of International Economics, 70*(1), 25–51.

Minca, C. (2000). 'The Bali Syndrome': The explosion and implosion of 'exotic' tourist spaces. *Tourism Geographies, 2*(4), 389–403.

Muryani, Permatasari, M. F., & Padilla, M. A. E. (2020). Determinants of tourism demand in Indonesia: A panel data analysis. *Tourism Analysis, 25*(1), 77–89.

Nguyen, C. P., Schinckus, C., & Su, T. D. (2020). Economic policy uncertainty and demand for international tourism: An empirical study. *Tourism Economics, 26*(8), 1415–1430.

Page, S. J. (2014). *Tourism management*. Abingdon, UK: Routledge.

Papatheodorou, A. (2000). *Accessibility and market structure: Implications for island tourism destinations in the Mediterranean Region*. Tourism on Islands and Specific Destinations. Paper. University of Surrey. UK. Retrieved from www.researchgate.net/publication/286232269_Accessibility_and_Market_Structure_Implications_for_Island_Tourism_Destinations_in_the_Mediterranean_Region

Park, S. H., Hsieh, C.-M., & McNally, R. (2010). Motivations and marketing drivers of Taiwanese island tourists: Comparing across Penghu, Taiwan and Phuket, Thailand. *Asia Pacific Journal of Tourism Research, 15*(3), 305–317.

Parra-López, E., & Martínez-González, J. A. (2018). Tourism research on island destinations: A review. *Tourism Review, 73*(2), 133–155.

Peters, M. (1980). The potential of the less-developed Caribbean countries. *International Journal of Tourism Management, 1*(1), 13–21.

Piermartini, R., & Rousová, L. (2013). The sky is not flat: How discriminatory is the access to international air services? *American Economic Journal: Economic Policy, 5*(3), 287–319.

Plog, S. (2001). Why destination areas rise and fall in popularity: An update of a Cornell Quarterly classic. *Cornell Hotel and Restaurant Administration Quarterly, 42*(3), 13–24.

Rodriguez, A. (2023, May 28, 2023). Hawai'i doesn't want visitors right now, or ever. Here's why. Retrieved from https://taiswim.co/blogs/bikini-blog/why-you-should-not-come-to-hawaii-for-your-next-vacation

Rosselló-Nadal, J. (2001). Forecasting turning points in international visitor arrivals in the Balearic Islands. *Tourism Economics, 7*(4), 365–380.

Shareef, R., & McAleer, M. (2008). Modelling international tourism demand and uncertainty in Maldives and Seychelles: A portfolio approach. *Mathematics and Computers in Simulation, 78*(2–3), 459–468.

Soh, A.-N., Puah, C.-H., & Arip, M. A. (2019). Construction of tourism cycle indicator: A signalling tool for tourism market dynamics. *Electronic Journal of Applied Statistical Analysis, 12*(2), 477–490.

State of Hawai'i (2021). *Visitor satisfaction and activity study: 2021 Annual Report*. Hawai'i: State of Hawai'i. Retrieved from www.hawaiitourismauthority.org/media/9940/2021_dbedt_vsat-annual-report-final.pdf

Takahashi, K. (2020). Comparing the determinants of tourism demand in Singapore and French Polynesia: applying the tourism demand model to panel data analysis. *Tourism Analysis, 25*(1), 175.

Uysal, M. (1998). The determinants of tourism demand: A theoretical perspective. In K. G. Debbage & D. Ioannides (Eds.), *The economic geography of the tourist industry* (pp. 79–98). London: Routledge.

Walker, T. B. (2020). A review of sustainability, tourism, and the marketing opportunity for adopting the Cittaslow model in Pacific small islands. *Tourism Review International, 23*(3–4), 99–114.

Yarde, K., & Jönsson, C. (2016). Implications for CARICOM member states in the search for a 'liberally controlled' air transport market: The case of regional tourism to Barbados. *Journal of Air Transport Management, 53*(June), 12–22.

Yiamjanya, S., & Wongleedee, K. (2014). International tourists' travel motivation by push-pull factors and the decision making for selecting Thailand as destination choice. *International Journal of Humanities and Social Sciences, 8*(5), 1348–1353.

Yun, J., Hao, S., Si, X., & Mei, W. (2023). Purchasing decision of tourism destination under tourism demand model-taking tourism statistics from Hawaii as example. *International Journal of Asian Social Science, 13*(2), 78–88.

3 Tourism policies to influence supply

3.1 Introduction

In many cases, islands are smaller land masses with smaller populations, and this situation results in resource constraints. Resources on islands started with mainstream agriculture or plantations, and with a decline in these markets changed to focus on tourism resources for economic growth (Parry & McElroy, 2009). Such a transition comes with challenges of supply as the required products are not produced on the island (Castellani & Sala, 2010). In the island of Mauritius, strategies to transition from an agriculture to tourism included giving attention to supply chain improvement (Gounder, 2022). Banking on tourism also resulted in a decline in agriculture and the importation of food supplies from other countries. A range of resources is needed to support a successful island tourism sector. The tourism industry is built on the availability of goods and services on island, and any shortages negatively affect operations. In addition, the supply of natural, human, and man-made resources faces several constraints. Thahir et al. (2021) have found that in the Togean Islands the key issues relating to the tourism supply chain include electricity, fishing, labour, licensing, roads, telecommunications, and waste.

Tourism products and services are created using resources. Resorts have been known to import virtually everything and the demand for importation increases for luxury items outside the country (Breiling, 2016). Visitors on island increase the day-to-day population and resources must be present to meet the visitors' needs. Differences in consumption practices between locals and tourists are important, and in many cases result in importation of goods. As a result, an efficient supply chain is necessary to facilitate successful tourism business operations. Gössling et al. (2012) have found that in islands with high visitor arrivals, water conflicts may occur. In the case of Mauritius, the proportion of tourism-related and domestic water consumption can be up to 40% of the domestic water supply (Gössling et al., 2012). Consumption of locally available food items is important for sustainable practices. Sustainable food consumption and production involve purchasing locally grown food (Linnes, Weinland, Ronzoni, Lema, & Agrusa, 2022). Sustainable food consumption on islands means the ready supply of locally available goods for food and beverage operations including restaurants.

DOI: 10.4324/9781003435112-4

Chapter 3 comprises three sections: island tourism supply, tourism supply chains, and policies for island tourism supply. The supply of resources on island faces several challenges and threatens island tourism sustainability. A tourism supply chain involves business relationships that facilitate the combination of goods and services including raw materials, which are comprised in finished products, and these are distributed and delivered to a final user, the tourist (Tapper & Font, 2004). A functional tourism supply chain is essential for island destinations. In that regard, policies that influence tourism supply in islands have been discussed.

3.2 Island tourism supply

Key resources are amalgamated to produce tourism products and services. Tourism supply elements can be related to a tourism system's micro-environment (Ritchie & Crouch, 2003). In reference to an island's micro-environment, Boukas and Chourides (2016) have argued that an island cannot sustain its competitive advantage in the long term because of limited comparative advantages and competition on a worldwide scale. Primarily, the agricultural and manufacturing sectors on an island are the main suppliers of produce and items for tourism purposes. Some examples of supplies for the tourism industry include raw and processed food, furniture, equipment, and consumables. In an island context, supplies are often constrained by the population size and must be replenished. Croes (2013) has suggested that islands specializing in tourism should allocate more resources to tourism than other sectors and by so doing increase supplies. In the context of a supply-side view of coastal tourism in Cyprus, Farmaki (2012) has noted the important government role of funding tourism promotion and low-interest loans to investors.

Entrepreneurship in island tourism requires greater attention. Porter, Orams, Lück, and Andreini (2022) have noted that residents of remote Filipino fishing communities lack understanding of market opportunities, which resulted in their non-participation in creating souvenirs from discarded shells. Many island tourism businesses are micro-, small-, and medium-sized enterprises (MSMEs), and women are employed in smaller businesses (Stephenson & Timothy, 2022). MSMEs that are women-owned and managed require additional support as women also have other responsibilities in households that require balancing tasks (Kimbu, Ngoasong, Adeola, & Afenyo-Agbe, 2019; Kokkranikal & Baum, 2011). Tourism businesses on islands face many challenges to survive and become successful with the help of financing, human resources, training and development, managerial skills and competencies, and marketing strategies. To improve profitability, businesses must engage in practices to reduce energy consumption, manage waste, and use recycled materials in managing tourism supply (Alkier, Milojica, & Roblek, 2022). The prospects are unfavourable for larger business enterprises as well. Large businesses are limited in islands because of the size of an island's population. In addition, large enterprises have growth constraints as it really depends on the success of the tourism sector in terms of visitor numbers. Nonetheless, the inclusion of trans-national corporations in small island destinations contributes to multiplier effects and allows countries to negotiate effectively with international tour operators (Barrowclough, 2007).

Intermediaries in a tourism system exert an extent of control over tourism supply. Parra López Eduardo and Baum (2004) have found that intermediaries exert control over the tourism offer and price in small island destinations. The situation is further exacerbated by buy outs by large operators and this practice reinforces foreign dominance (Parra López Eduardo & Baum, 2004). Marketing intermediaries facilitate and control the demand and supply nexus. Sigala (2008) has noted that tour operators can promote sustainable tourism development through their role to direct tourists to destinations and suppliers. Intermediaries balance out the use of tourism resources by adjusting bookings and are essential players in the tourism industry because of resource utilization. In the process of performing an intermediary role, local providers may have limited control on the results from this process. Added to this, local tourism businesses are unable to successfully nego-tiate deals (Nyakunu, 2014). As a result, marketing intermediaries, that are largely foreign based, hold the greater power to control the use of resources in island destinations. For example, intermediaries represent various types of tourists, con-trol the flow of tourists, and the utilization of resources. In addition, all-inclusive products are provided through foreign tour operators, and restrict the supply of local products to all-inclusive tourists (Farmaki, Altinay, & Yaşarata, 2016).

Ownership structures of the related and supporting industries are important to facilitate tourism activities. Community ownership of tourism supply provision can assist with supporting sustainability practices (Roxas, Rivera, & Gutierrez, 2020). Ownership of airlift through subsidies or direct ownership can help mitigate vul-nerabilities and risks in island tourism (Caribbean Tourism Organization, 2020). Ownership of parks and recreational areas, whether public or private sector owner-ship, comes with its own challenges. The upkeep of the resources requires fees and charges that go towards maintenance and staffing, and event-related activities also require coordination and management (Tapper & Font, 2004).

3.3 Island tourism supply chains

The multiplicity of tourism products and services warrants an in-depth study of supply chains on islands. Supply chain management facilitates the growth in tourism demand (Alkier et al., 2022). First, many small islands import items required for tourism operations. Parra López Eduardo and Baum (2004) have noted the loss of foreign currency because of importing goods for tourism in the Canary Islands. Second, depending on the location of an island, supplies may take weeks to arrive by sea as air transport is too expensive. Mandal, Roy, and Raju (2016) have argued for using a resource-based view to develop tourism supply chain agility that will result in easy resource deployment to increase com-petitive advantage. Third, the cost of transporting certain products makes the tourism product expensive. Bearing these circumstances in mind, the logistics of supply chain management is complicated in islands. Knowledge of government regulations from the point of departure to the point of arrival is important to ensure smooth processing of goods. Filling skill gaps and shortages in islands are critically important for supply chain management, and skill gaps and shortages affect the retention of trained staff and levels of service delivery and quality (Baum, 2018; Carlsen & Butler, 2011).

Supply chains to islands were disrupted during the last global pandemic as shortages in supply of goods were exacerbated by the shut down in international transport, and shipping price increases (Campbell & Connell, 2021; Gounder, 2022). Supply chain processes are activated by functional inventory systems, which account for the day-to-day demands from customers, and such systems are needed to support smooth operations. Once imported goods arrive, these must be distributed to a consumption point. Adequate and suitable transport modes are then needed to ensure the goods are delivered with the required quality in a timely manner. In some islands refrigerated trucks have been an issue, and therefore perishable items deteriorate quickly. Inadequate cold storage units and trucks in the Seychelles and Jamaica create supply chain dysfunction (Hampton & Jeyacheya, 2013). Inadequate land transport has two problems: increasing the cost to obtain new items, and creating customer dissatisfaction as the product cannot be consumed. Added to this, imported goods generate waste for islands that have inadequate waste management infrastructure. Diaz-Farina, Díaz-Hernández, and Padrón-Fumero (2020) have noted the need to institute charges for mixed waste to encourage recycling throughout the tourism supply chain in Tenerife.

Local resource usage supports the effective functioning of tourism supply chains on islands. The involvement of local suppliers in tourism supply depends on the scale of the businesses in the tourism sector. As a result, one policy approach is the creation of linkage policies to support the supply of goods by local suppliers to tourism businesses, and such a policy could assist with poverty reduction. In the case of Mauritius, Gounder (2022) has recommended policies that support employment-led growth and capital formation to improve the value addition and supply chains in tourism. Another important resource to support particularly the supply of agricultural goods is land. The distribution of land resources and land tenure policies, for tourism versus other land use, are important to provide equitable use practices. In the case of Barbuda, a communal system of land tenure has hindered tourism development (Weaver, 1998), and there are implications for the local supply chain. Islands have limited land space, and use of rich arable lands for tourism could undermine efforts to create a viable agricultural sector to support tourism in a way to maximize benefits from the sector and reduce the cost of doing the business of tourism on an island. The supply of water on small islands has become increasingly challenging. A case study about Bali and the water crisis follows.

Mini Case Study 2: Tourism in Bali is drying the island

Indonesia is the largest archipelago of islands in the world. The country's tourist arrivals growth has been steadily increasing since 2009 from 6.3 to 16.1 million visitors in 2019 (Figure 3.1). Bali is an Indonesian island (Bali Tourism Board, 2023). The economy on the island of Bali is dominated by tourism and agriculture. With many small businesses supporting the tourism sector, the local economy benefits from the tourism industry (Bali Tourism Board, 2023). Amid a vibrant tourism sector, Bali is facing a water crisis that has affected the tourism sector and supporting industries such as agriculture (Milko & Jatmiko, 2022; Ritter, 2019). With a water system that supports both the residents and tourists dating back in the 9th century, and excessive pumping to meet water needs, salt water has entered the water table (Ritter, 2019). As an island in an archipelago, Bali has high numbers of both international and domestic tourists, with 2019 estimates of 6.2 million and 10.5 million, respectively, and an island population of 4.3 million (Milko & Jatmiko, 2022). Cole (2012) has pointed out that the inequity of water supply in Bali has resulted in poorer communities' lack of water access and recommended positive change with greater enforcement of rules and law.

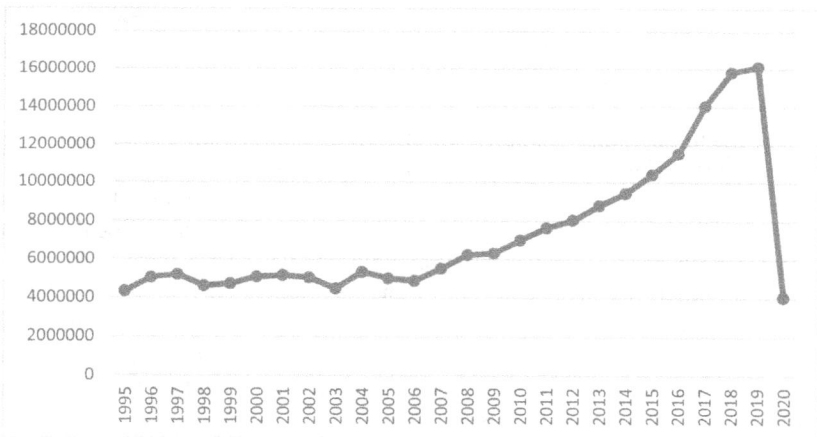

Figure 3.1 Tourist arrivals growth in Indonesia (1995–2020).

Source: World Bank (2023).

Mini Case Study 2 discussion questions

1 Explain how a lack of water supply affects living on an island with a dominant tourism industry and one without a dominant tourism industry.
2 What are the policies that may be instituted to improve the supply and distribution of water in Bali?
3 Discuss the sustainability of water resources in island destinations.

3.4 Policies for island tourism supply

A resource-based policy framework for island tourism supply considers the key policies that support the provision of tourism products and services. A tourism sector pulls resources from several directions resulting in tourism impacts (Mason, 2020). Transportation services are critical for delivery of tourism experiences throughout various points on an island. Using the Balearic Islands as a case study, Aguiló, Palmer, and Rosselló (2012) have considered a taxation policy to regulate the trade-off between the use of different modes of transport by the various tourist market segments. The road network, parking, and access to sites and attractions are also important aspects of the transportation system to furnish tourism supplies. Baldacchino and Spears (2007) have analysed the bridge effect of creating a 'fixed link' between the mainland and Prince Edward Island. While additional transportation links support tourism supply, other effects such as making the island less exclusive and increase in property prices may occur (Baldacchino & Spears, 2007). Consideration of whether an island serves as a home port for cruise ships will assist with providing for access at the various cruise ports of entry for distribution trucks. Homeporting at Caribbean ports was suggested based on a no-sail order from Florida ports during the last global pandemic (Ajagunna & Casanova, 2022). Although homeporting expands the economic impact of tourism, the various regulations and codes must be enforced to ensure health and safety requirements are met.

The water supply on islands has received increasing attention as islands grapple with desertification. The provision of fresh water sources for use by the local population and for tourism purposes is becoming increasingly scarce on island destinations. Island tourism, in the Mediterranean, with a warmer climate and low rainfall contributes to water scarcity as water easily evaporates (Essex, Kent, & Newnham, 2004). Carlsen and Butler (2011) have noted in Lakshadweep, a shortage of freshwater and inadequate desalination plant provision, and as a result, limit tourism activities on the islands. Water shortages are problematic on islands and in the case of the Togean Islands, in Indonesia, water has to be imported from one island to the next, and water importation increases the operational cost of resorts (Thahir et al., 2021). Issues about rainfall collection, process and distribution of water requires greater attention by policymakers. The wastage of water through use of poor, degraded infrastructure is also a concern. Gössling et al. (2012) have recommended that tourism water policies include carbon reduction, building codes, monitoring, and charging for water consumption.

Education and environmental management policies support the supply of tourism products and services (Alkier et al., 2022). Supply of sustainable food creates carbon and water footprints, and the range of education and environmentally friendly policies address concerns about sustainable food practices (Linnes et al., 2022). The use of the natural environment should be equally balanced with conservation practices for future use. Management of waste and wastewater threaten the fragile land and marine ecosystems of islands. Disease control, public health, safe and hygienic practices, and environments ensure that hosts' and guests' health are maintained. Water resource management on islands can improve water supply, wastewater treatment, and law enforcement (Phong & Van Tien, 2021).

Climate change adaptation affects tourism supply in several ways. First, the use of renewable energy comes with a cost to tourism businesses and a balance must be found in the most effective ways to manage these renewable resources on islands (Becken & Kaur, 2021). Second, climate change impacts such as rising sea levels greatly impact developments along the coastline that are popular tourism business clusters (Moghal & O'Connell, 2018). Third, islands across the globe have faced many natural disasters that have shut down tourism plants for varying lengths of time (McLeod, 2022). Breiling (2016) has pointed out the need for disaster risk reduction, particularly considering the effects on the tourism supply chain including shortages of fuel, food stuff and supplies, and risks in water supply. The articulation of a set of climate change adaptation policies to address tourism supply in islands is critically needed. Klint et al. (2012), using the case of Vanuatu, have suggested that more explicit climate change adaptation policies are needed for island tourism.

Financial resources are needed to develop a tourism plant. During the recent global pandemic it was imperative to institute a range of policies including, fiscal policies of SME, liquidity, loan, subvention and government administration, and monetary policies of employment, tax, interest, investment, and bank, to address the contraction in tourism demand and support tourism suppliers (Şengel, Işkın, Çevrimkaya, & Genç, 2022). Policies that incentivize and facilitate the supply of tourism products and services include tax holidays, tax breaks, and preferential loan rates from domestic banks to name a few. Local business financial support is essential to facilitate local tourism supply. Offering local businesses credit lines allow competition with foreign counterparts (Baidal, 2003). The requirement to include a local investor as part of the development package facilities further development of local knowledge and assists a foreign investor with navigating the local landscape. In addition, funds that include islands have been established to address challenges with climate adaptation (Klint et al., 2012).

A resource-based policy framework for island tourism supply has been proposed. The framework addresses transportation, water, education, climate adaptation, and financial policies for island tourism (Figure 3.2). Island tourism supply policies start with climate adaptation policies followed by water management policies. Addressing the challenges of the effects of climate change including risks of natural disaster and changes to island landscapes is a number one priority. In one case, the people of Kiribati, with a population about 113,000, purchased climate-refuge land in Fiji in 2014, and have been working the land to support the Kiribati (Reed, 2023). Water policies, also connected with climate adaptation from the perspective of rainfall and rising temperatures, are important for the sustainability of island tourism. Education, transportation, and financial policies are supporting policies for effective and efficient tourism supply (Figure 3.2).

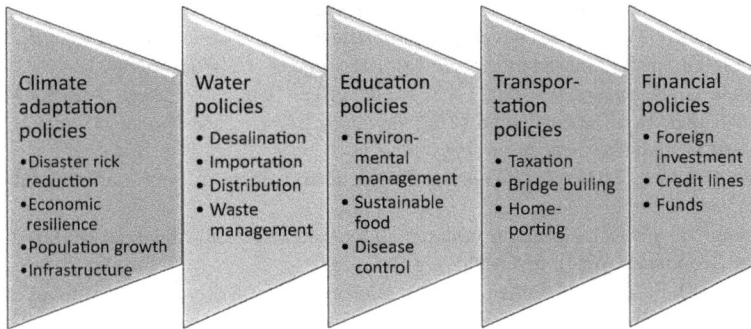

Figure 3.2 Island tourism supply resource-based framework.

3.5 Conclusion

This chapter discusses the characteristics of tourism supply on islands. The operation of a successful tourism sector has resource constraints including limited land, capital, human and financial resources. A range of policies is needed to counteract the challenges that islands face to facilitate tourism supply. Local supply and distribution chains, environmental policies, financial policies, water policies, and the matter of climate adaptation policies were discussed. A resource-based policy framework for island tourism supply contributes to understanding prioritization of policy initiatives in the sustainable development of islands. A mini case study about the water crisis in Bali is included to explore the sustainability of island tourism.

Chapter 3 discussion questions

1 Construct and explain a diagram of tourism supply issues in a remote island.
2 What are the key considerations in developing a successful tourism supply chain in a developing country island destination?
3 Apply and discuss the island tourism supply resource-based framework in an island that has declining visitor numbers in the last 5 years and an island that has increasing visitor numbers in the last 5 years.

References

Aguiló, E., Palmer, T., & Rosselló, J. (2012). Road transport for tourism: Evaluating policy measures from consumer profiles. *Tourism Economics, 18*(2), 281–293.

Ajagunna, I., & Casanova, S. (2022). An analysis of the post-COVID-19 cruise industry: Could this be a new possibility for the luxury yacht sector in the Caribbean? *Worldwide Hospitality and Tourism Themes, 14*(2), 115–123.

Alkier, R., Milojica, V., & Roblek, V. (2022). The complexity of the tourism supply chain in the 21st century: A bibliometric analysis. *Kybernetes, 52*(11), 5480–5502.

Baidal, J. A. I. (2003). Regional development policies: An assessment of their evolution and effects on the Spanish tourist model. *Tourism Management, 24*(6), 655–663.

Baldacchino, G., & Spears, A. (2007). The bridge effect: A tentative score sheet for Prince Edward Island. In G. Baldacchino (Ed.), *Bridging islands: The impact of "fixed links"* (pp. 49–68). Charlottetown: Acorn Press.

Bali Tourism Board (2023). Bali Tourism Board. Retrieved from www.balitourismboard. org/stat_arrival.html

Barrowclough, D. (2007). Foreign investment in tourism and small island developing states. *Tourism Economics, 13*(4), 615–638.

Baum, T. (2018). Sustainable human resource management as a driver in tourism policy and planning: A serious sin of omission? *Journal of Sustainable Tourism, 26*(6), 873–889.

Becken, S., & Kaur, J. (2021). Anchoring "tourism value" within a regenerative tourism paradigm – A government perspective. *Journal of Sustainable Tourism, 30*(1), 52–68.

Boukas, N., & Chourides, P. (2016). Niche tourism in Cyprus: Conceptualising the importance of social entrepreneurship for the sustainable development of islands. *International Journal of Leisure and Tourism Marketing, 5*(1), 26–43.

Breiling, M. (2016). *Tourism supply chains and natural disasters: The vulnerability challenge and business continuity models for ASEAN countries.* Jakarta: Economic Research Institute for ASEAN and East Asia.

Campbell, Y., & Connell, J. (2021). Introduction: COVID-19 and small island states. In Y. Campbell & J. Connell (Eds.), *COVID in the islands: A comparative perspective on the Caribbean and the Pacific* (pp. 1–39). Singapore: Palgrave Macmillan.

Caribbean Tourism Organization (2020). Caribbean sustainable tourism policy and development framework. Retrieved from Barbados: https://ourtourism.onecaribbean.org/resour ces/caribbean-sustainable-tourism-policy-framework-2020/

Carlsen, J., & Butler, R. (2011). Introducing sustainable perspectives of island tourism. In J. Carlsen & R. Butler (Eds.), *Island tourism: Sustainable perspectives* (pp. 1–7). Wallingford, UK: CABI.

Castellani, V., & Sala, S. (2010). Sustainable performance index for tourism policy development. *Tourism Management, 31*(6), 871–880.

Cole, S. (2012). A political ecology of water equity and tourism: A case study from Bali. *Annals of Tourism Research, 39*(2), 1221–1241.

Croes, R. (2013). Tourism specialization and economic output in small islands. *Tourism Review, 68*(4), 34–48.

Diaz-Farina, E., Díaz-Hernández, J. J., & Padrón-Fumero, N. (2020). The contribution of tourism to municipal solid waste generation: A mixed demand–supply approach on the island of Tenerife. *Waste Management, 102*(February), 587–597.

Essex, S., Kent, M., & Newnham, R. (2004). Tourism development in Mallorca: Is water supply a constraint? *Journal of Sustainable Tourism, 12*(1), 4–28.

Farmaki, A. (2012). A supply-side evaluation of coastal tourism diversification: The case of Cyprus. *Tourism Planning & Development, 9*(2), 183–203.

Farmaki, A., Altinay, L., & Yaşarata, M. (2016). Rhetoric versus the realities of sustainable tourism: The case of Cyprus. In P. Modica & M. Uysal (Eds.), *Sustainable island tourism: Competitiveness, and quality-of-life* (pp. 35–50). Wallingford, UK: CABI.

Gössling, S., Peeters, P., Hall, C. M., Ceron, J.-P., Dubois, G., & Scott, D. (2012). Tourism and water use: Supply, demand, and security. An international review. *Tourism Management, 33*(1), 1–15.

Gounder, R. (2022). Tourism-led and economic-driven nexus in Mauritius: Spillovers and inclusive development policies in the case of an African nation. *Tourism Economics, 28*(4), 1040–1058.

Hampton, M., & Jeyacheya, J. (2013). *Tourism and inclusive growth in small island developing states*. London: Commonwealth Secretariat.

Kimbu, A., Ngoasong, M., Adeola, O., & Afenyo-Agbe, E. (2019). Collaborative networks for sustainable human capital management in women's tourism entrepreneurship: The role of tourism policy. *Tourism Planning & Development, 16*(2), 161–178.

Klint, L. M., Wong, E., Jiang, M., Delacy, T., Harrison, D., & Dominey-Howes, D. (2012). Climate change adaptation in the Pacific Island tourism sector: Analysing the policy environment in Vanuatu. *Current Issues in Tourism, 15*(3), 247–274.

Kokkranikal, J., & Baum, T. (2011). Tourism and sustainability in the Lakshadweep Islands. In J. Carlsen & R. Butler (Eds.), *Island tourism: Sustainable perspectives* (pp. 54–71). Wallingford, UK: CABI.

Linnes, C., Weinland, J. T., Ronzoni, G., Lema, J., & Agrusa, J. (2022). The local food supply, willingness to pay and the sustainability of an island destination. *Journal of Hospitality and Tourism Insights, 6*(3), 1328–1356.

Mandal, S., Roy, S., & Raju, G. A. (2016). Tourism supply chain agility: An empirical examination using resource-based view. *International Journal of Business Forecasting and Marketing Intelligence, 2*(2), 151–173.

Mason, P. (2020). *Tourism impacts, planning and management*. Abingdon: Routledge.

McLeod, M. (2022). Tourism destination recovery, a case study of Grand Bahama Island. In I. Bethell-Bennett, S. Rolle, J. Minnis, & F. Okumus (Eds.), *Pandemics, disasters, sustainability, tourism* (pp. 93–108). Leeds, UK: Emerald Publishing.

Milko, V., & Jatmiko, A. (2022, December 13, 2022). *Bali's water crisis threatens local culture, UNESCO sites*. New York: The Associated Press.

Moghal, Z., & O'Connell, E. (2018). Multiple stressors impacting a small island tourism destination-community: A nested vulnerability assessment of Oistins, Barbados. *Tourism Management Perspectives, 26*(April), 78–88.

Nyakunu, E. P. (2014). *The role of small and medium enterprises (SMEs) in tourism policy formulation in Namibia*. University of Johannesburg, South Africa.

Parra López Eduardo, D., & Baum, T. (2004). An analysis of supply-side relationships in small island destinations: The role of tour operators, travel agencies and tourism transport in the Canary islands. *Tourism and Hospitality Planning & Development, 1*(3), 201–218.

Parry, C. E., & McElroy, J. L. (2009). The supply determinants of small island tourist economies. *The ARA (Caribbean) Journal of Tourism Research, 2*(1), 13–22.

Phong, N. T., & Van Tien, H. (2021). Water resource management and island tourism development: Insights from Phu Quoc, Kien Giang, Vietnam. *Environment, Development and Sustainability, 23*(12), 17835–17856.

Porter, B., Orams, M. B., Lück, M., & Andreini, E. M. (2022). Trash or treasure? A qualitative exploration of gleaning by-products in tourism supply chains in remote Filipino Fishing Communities. *Advances in Southeast Asian Studies, 15*(1), 87–102.

Reed, B. (2023, February 23, 2021). Kiribati and China to develop former climate-refuge land in Fiji. Retrieved from www.theguardian.com/world/2021/feb/24/kiribati-and-china-to-develop-former-climate-refuge-land-in-fiji.

Ritchie, J. B., & Crouch, G. I. (2003). *The competitive destination: A sustainable tourism perspective*. Wallingford, UK: CABI.

Ritter, K. (2019). HotSpots H$_2$O: Rivers run dry in Bali as tourism, drought, overwhelm water supply. Retrieved from www.circleofblue.org/2019/hotspots/hotspots-h2o-rivers-run-dry-in-bali-as-tourism-drought-overwhelm-water-supply/

Roxas, F. M. Y., Rivera, J. P. R., & Gutierrez, E. L. M. (2020). Framework for creating sustainable tourism using systems thinking. *Current Issues in Tourism, 23*(3), 280–296.

Şengel, Ü., Işkın, M., Çevrimkaya, M., & Genç, G. (2022). Fiscal and monetary policies supporting the tourism industry during COVID-19. *Journal of Hospitality and Tourism Insights, 6*(4), 1485–1501.

Sigala, M. (2008). A supply chain management approach for investigating the role of tour operators on sustainable tourism: The case of TUI. *Journal of Cleaner Production, 16*(15), 1589–1599.

Stephenson, M. L., & Timothy, D. J. (2022). Future research trajectories: Pacific Island tourism. In M. L. Stephenson (Ed.), *Routledge Handbook on tourism and small island states in the Pacific* (pp. 381–392). Abingdon: Routledge.

Tapper, R., & Font, X. (2004). Tourism supply chains. *Report of a desk research project for the travel foundation, 23*. Leeds: Leeds Metropolitan University.

Thahir, H., Hadi, S., Zahra, F., Arif, I., Murad, M. A., & Lolo, M. H. (2021). Issues, challenges and strengths of sustainable tourism supply chain after Covid-19 in Togean National Park-Sulawesi, Indonesia: A preliminary findings. *Advances in Economics, Business and Management Research, 163*, 274–278.

Weaver, D. B. (1998). Peripheries of the periphery: Tourism in Tobago and Barbuda. *Annals of Tourism Research, 25*(2), 292–313.

World Bank (2023). World development indicators data bank. Retrieved from https://datab ank.worldbank.org/source/world-development-indicators

4 Tourism policies supporting sustainability

4.1 Introduction

Island tourism depends on island sustainability or is it the other way around. To the extent that a tourism sector supports government and other economic, socio-cultural, and environmental activities, the tourism activities contribute to island sustainability. To the extent that an island is developed in a sustainable manner, meaning, minimizing the negative consequences of development, the tourism activities can be sustained. Sustainability practices on islands are at the root of building a sustainable island destination. Such practices are supported by a range of policies that may or may not be instituted by a tourism policy framework. Policies to influence tourism supply (Alkier, Milojica, & Roblek, 2022), incentives to adopt sustainability practices (Farmaki, Altinay, & Yaşarata, 2016), and green hushing to selectively communicate sustainability practices in businesses (Font, Elgammal, & Lamond, 2017) contribute to sustainability outcomes in the tourism sector in positive or negative ways. In addition, departments with responsibility for planning, economic, transportation, and infrastructure activities are key to provide the supportive mechanisms for tourism development. Adrianto et al. (2021) have supported incorporating carrying capacity in tourism development models in the Tidung Islands of Indonesia to improve sustainable tourism. Policies that affect the tourism sector span a range of government departments and this complicates island tourism policy development for sustainability. As part of any successful tourism destination is an understanding of both the policy content that supports sustainability and the processes that bring about optimal tourism policies to result in island sustainability.

An important consideration for island governments is whether to embark on a tourism development path. Some authors view the tourism sector as being fragile and unsustainable because of the high dependence on foreign inputs and lack of economic diversification (Bernard & Cook, 2015; Boukas & Ziakas, 2016; Grilli, Tyllianakis, Luisetti, Ferrini, & Turner, 2021). Other authors propose a tourism-led growth hypothesis (TLGH) (Brida, Cortes-Jimenez, & Pulina, 2016). At the foundation of a TLGH, tourism supply is associated with economic growth and funnelled by tourism demand. Schubert, Brida, and Risso (2011) have found that

DOI: 10.4324/9781003435112-5

in Antigua and Barbuda during the period of 1970–2008, tourism demand resulted in economic growth.

Policies in several key areas are needed for the development of a successful island destination. Foreign direct investment (FDI) expands tourism capacity. Fauzel, Seetanah, and Sannassee (2017) have provided an explanation of the importance of FDI in Mauritius as with FDI flows to the non-tourism sectors, spill-over effects on the tourism sector were evident. Increasing the number of visitors further increased FDI in the tourism sector (Fauzel et al., 2017). In addition, Hambira and Saarinen (2015) have pointed out that the tourism sector is lagging in climate change adaptation policies. Becken, Whittlesea, Loehr, and Scott (2020) have recommended policy integration between the climate change and tourism domains to address tourism policy gaps. Several island states are comprised in archipelagos. Tourism policies for islands and archipelagos require specific consideration of sustainable development, including land usage and zoning strategies, taxation, accommodation, immigration, planning, transport, and intraregional tourism policies (Bardolet & Sheldon, 2008). This chapter includes three sections. The first section considers tourism growth and de-growth in islands, the second section reviews sustainability goals and policy formulation, and the third section outlines implementation of policy changes. Islands contend with massive influxes of tourists beyond the coping capacity of island infrastructure. Given this circumstance, this chapter sets out policies to support sustainability of islands as tourism destinations.

4.2 Island tourism growth and de-growth

Sustainable development of islands suggests that tourism growth happens in a manner that will optimize positive effects of growth (Lim & Cooper, 2009). Policies that support sustainable development within the context of tourism growth and de-growth are important for island sustainability. Tourism growth may be viewed as either being controlled or uncontrolled. Page (2014) has presented two concepts about tourism growth. The first concept is the traditional one and is referred to as the snowball concept that results in rapid, uncontrollable, overdeveloped, and unattractive tourism growth (Page, 2014). In island tourism economies, mass tourism is seen as uncontrolled growth (Xing & Dangerfield, 2018). Bernardo and Jorge (2019) have noted that uncontrolled growth affected residents and is perceived more by younger age groups than older age groups, and more highly educated respondents. Uncontrolled growth is viewed as being exploitative and makes the destination unattractive (Castellani & Sala, 2010). The problems of uncontrolled growth are not surface deep, and are more related with the inadequate infrastructure to support the rapid tourism growth and as a result the unsatisfactory residents' quality of life (Gill, 2004).

The second concept about tourism growth relates to controlled growth. A controlled growth concept is similar to an amoeba concept with tourism growth starting at a single-cell reproducing itself, and the cells adapt and change as growth continues (Page, 2014), bringing controlled growth into balance. The concept of controlled growth must be supported by facilitating policies that result in sustainable development. A sustainable development path for tourism must envisage control of tourism growth. Uncontrolled growth can be associated with mass tourism, whereas controlled growth can be associated with alternative or non-mass tourism such as ecotourism, and culture and heritage tourism. Alternative tourism is small scale with less numbers of visitors to the destination (Mason, 2020). Although Aruba has articulated controlled growth as a policy, Peterson (2020) has pointed out that the island has reached its optimum growth. Controlled growth through development of alternative tourism contributes to positive impacts on host communities by cross-cultural interaction between hosts and visitors (Pratt, Gibson, & Movono, 2013).

Moving away from uncontrolled tourism growth is supported through policy interventions. The measurement of tourist arrivals is the most common means to understand tourism growth (Butler, 1980). A decline in tourist arrivals can affect economic growth in islands (Kumar & Patel, 2023), and a way has to be found to transition island tourism to sustainable island tourism (Telesford, 2022). For island tourism sustainability, the ongoing supply of goods and services is critical. Considerations from a multilevel perspective (MLP) about the fulfilment of tourism products and services enhances policies regarding constructing large resorts along the coastline that are highly dependent on goods and services from overseas (Telesford, 2022). Sustainable tourism on islands requires deconstructing and reconstructing of the accommodation sector. Decoupling the impetus of mass tourism and de-growing tourism in islands require changes in tourism policies. A policy change that allows the purchase of locally produced supply has been proposed by Walker, Lee, and Li (2021).

4.3 Island tourism policies: goals and formulation

The complexity of tourism policy formulation results in policy emergence and decline (Stevenson, Airey, & Miller, 2009), and this means that adjustments are constantly being made to formulate effective tourism policies. A starting point for tourism policy formulation is goal-setting. A policy framework that outlines policy instruments and indicators is important towards achievement of tourism goals (Hall, 2011). Tourism goal setting is evident within the tourism planning and strategic management activities. Tourism plans seek to provide sustainable and competitive tourism destinations. Within that direction an integrated, monitored long-term plan, at an appropriate pace, that involves all stakeholders, available knowledge, bearing in mind the cost and carrying capacity limits, may be achievable (Castellani & Sala, 2010). Plans are focused but goals need to be carefully defined. Goals are more abstract and should be defined based on specific objectives and strategies (Farsari, 2012). Tourism policies may be objective driven to increase international tourists and promote overseas travel to locals (Zhang, Chong, & Ap, 1999). During the COVID-19 pandemic, The Bahamas set out specific goals relating to the implementation of point of entry best practices and reopening of domestic borders and later on demand stimulation (McLeod, 2021). Although there may be different objectives, these are stated to achieve common goals (Wisnumurti, Darma, & Putra, 2021).

Tourism policies are selected for tourism planning and development. Edgell (1987) has developed a framework for tourism policy formulation that has summarized important policy concerns under economic, environmental, and socio-cultural issues. Gruetzmacher (2021) has analysed sustainable tourism policies in nine island countries and has found four policy categories including direct and/or quantifiable tourism management, management of tourist behaviour and types, integration of local needs, and indirect contributions to sustainable tourism. Another approach to achieve sustainable tourism is to set limits. Limits to tourism growth is a planning policy adopted to ensure sustainability (Mai & Smith, 2015). In the context of the global pandemic in islands, McLeod (2021) has found that tourism policies relating to tourism demand, promotion, travel safety, and risk were particularly needed. Policies can be developed from critical issues. For example, the top ten issues facing tourism include the top three issues relating to the recovery from the global pandemic, particularly regaining economic confidence in tourism as a sector in the economy and modes of transportation to facilitate tourism activities (Edgell, 2022). Supporting innovation is another policy direction. Hall (2009) has considered the intersection between innovation policy and tourism policy and concluded that tourism has not gained recognition as an innovative industry. Four types of innovations were identified including product, process, organizational, and marketing innovations (Hall, 2009).

Tourism policy responses to climate change are critical for island tourism sustainable development. Belle and Bramwell (2005) have found in the case of Barbados, increasing public awareness supports potentially expensive strategies of climate mitigation. Tandrayen-Ragoobur and Fauzel (2021) have examined the effects of climate change and tourism governance on economic growth in 19 small island developing states (SIDS) over 23 years. While climate change and environmental changes affect economic growth, changes in governance are proposed to handle the impacts of climate change (Tandrayen-Ragoobur & Fauzel, 2021). Policies about climate change adaptation must be set out to achieve sustainability in the tourism sector.

Specific policies must be formulated in keeping with defined goals. Policy formulation is a political process that involves relationships of power, regulatory and legislative frameworks. Policy formulation occurs within policy networks through formal interactions of tourism stakeholders on boards, reporting and information sharing relationships (McLeod, 2023; McLeod, Chambers, & Airey, 2018). Farsari (2012) has set out the interrelationships of policy issues around sustainable tourism and noted that policy issues are similar at the local, regional, and national level, however, policy issues may be categorized according to those most influenced by other issues and those with the most influence on sustainable tourism. The achievement of policy goals involves a political process to set up policy priorities and achieve a shared common purpose (Nyakunu, 2014). Party politics may derail tourism policy formulation and implementation and therefore legislative frameworks that withstand party politics are preferred (Dela Santa & Saporsantos, 2016). A case study about sustainable tourism in Réunion Island follows.

Mini Case Study 3: Réunion Island remaining green and pristine

According to the Réunion Island Declaration on Sustainable Tourism, tourism accounts for 40% of the exports in about half small island developing states, and tourism development brings considerable challenges for the availability of fresh water, energy, and waste management (United Nations, 2013). Réunion Island is a French overseas territory located in the Indian Ocean and has a volcanic geomorphology. Since the 1980s, generation of energy is based on burning a by-product of sugar cane, dams, wind turbines, and solar panels (Petit Futé, 2023). Sugar cane is the island's main export product and accounts for 58% of cultivated land (Petit Futé, 2023).

With a population of over 800,000, the island's urbanization process has resulted in road and traffic congestion (Baddour & Percebois, 2012). Tourism infrastructure on Réunion Island benefits from the investment of trans-national corporations (Barrowclough, 2007), however, the island's remoteness makes it an expensive place to live as most items are imported from France (Petit Futé, 2023). Charles, Darné, and Hoarau (2019) have identified that stagnation in tourist arrivals since the beginning of the 2000s is because of endogenous impediments and unsuitable public policies rather than exogenous shocks. One way to counteract this is an expansion of tourism activities facilitated through new hotels, new air lift from emerging source markets, and a clear tourism vision (Charles et al., 2019).

Mini Case Study 3 discussion questions

1 Identify and categorize the main tourism policy issues in Réunion Island.
2 Analyse five policy goals to achieve sustainable island tourism.
3 Discuss the challenges of island tourism formulation and implementation on Réunion Island.

4.4 Island tourism policies: changes and implementation

Given the resource constraints of island economies, prioritization and selective allocation of resources are needed to result in effective and efficient policy implementation. While governments are the key drivers of tourism policy changes, changes in tourism policy contents require accurate data (Hassan & Burns, 2014). Additionally, governance shifts at the local and regional levels could bring about tourism policy changes (Church, 2004). Policy changes such as greater community involvement in tourism has had mixed results (Church, 2004). A mass tourism strategy will result in policy changes as such a strategy focuses on revenue maximization (Garcia, 2014). García (2013) has found that mass tourism resulted in changes in Spanish tourism policy, strong regional polarization, and territorial imbalances. Changes may come from external actors. Jenkins (2015) has argued foreign influences limit developing countries' power to effect tourism development. Added to this, the force of change is slowed by access to development funds, an additional difficulty for tourism development in the long run (Jenkins, 2015). Financial challenges may drive policy changes. Based on certain tax or visa requirements tourism policies may change (Dwyer, 2015). Changes in government funding may also drive policy changes. A devolution of funding from a central tourism authority to decentralized bodies changes the role of the agencies and changes tourism policy contexts (Richards, 1995). Across country borders within a region such as Europe, tourism policy coherence is a must to partake in future growth opportunities for the tourism sector (Lickorish, 1991). Changes in market conditions require fast facilitating actions, and supporting tourism policies must be created.

Economic and sustainable development policy changes occur. Figure 4.1 illustrates key tourism policies that support the sustainability pillars of economic, socio-cultural, and environmental activities. Economic policies that improve tourism supply are needed (Charles et al., 2019) and include employment creation, taxation, and poverty alleviation (Dwyer, 2015). Socio-cultural policies are needed to support tourism development on islands and the interaction between host and visitor. Community involvement spreads tourism benefits and minimize negative tourism impacts (Wisnumurti et al., 2021). Community empowerment in tourism development enhances economic linkages through job creation (Walker et al., 2021). Environmental policies based on using computable general equilibrium modelling of the impacts of climate-related variables have started to emerge (Dwyer, 2015). Belle and Bramwell (2005) have found that policy changes relating to policy responses to climate change require increasing public awareness and

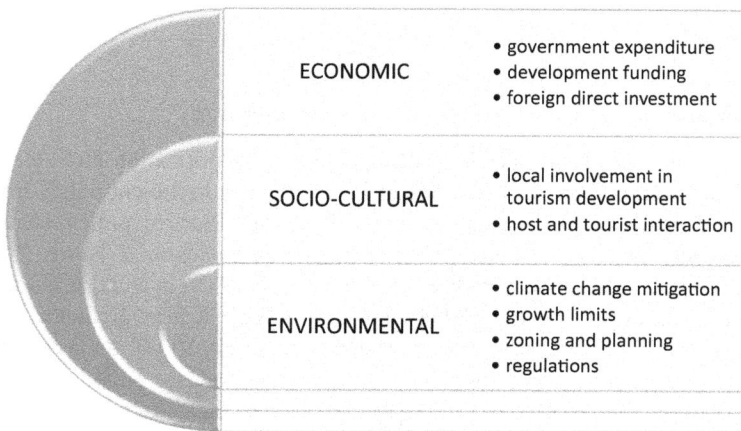

Figure 4.1 Tourism policies supporting sustainability.

policy formulation. Key perceptions from policymakers about climate change impacts include sea-level rise, beach alterations, damage to coastal tourism facilities and marine ecosystems (Belle & Bramwell, 2005). Policy changes that will address these impacts are therefore warranted and should be developed. Economic public policy changes affect tourism policy. Bohlin, Brandt, and Elbe (2014) have pointed out that a shift from welfare provision to competition resulted in tourism policy changes and that regulations substantially contribute to changing tourism policy.

4.5 Conclusion

Sustainability of an island tourism sector is dependent on the growth path set out by tourism stakeholders. Outward-oriented growth with high imports of goods and services limits sustainable tourism in islands as a cut-off from supplies means that tourism activities are not supported. To avoid this circumstance an inward-oriented growth path that is supported by local businesses is preferred. To set out a vision for tourism development in islands, one must consider the resources that are available to sustainably support the tourism sector. Indigenous resources that are home-grown can create unique tourism products and services for visitors and should be cultivated. Tourism models that create dependency of the tourism sector on imports are to be avoided if islands are to become self-sustaining tourism hubs.

Chapter 4 discussion questions

1 Analyse five goals to resolve issues about climate change in island destinations.
2 Discuss the statement that not all policies are effective in the achievement of sustainability.
3 Apply the concept of environmental sustainability to a mass tourism island destination.

References

Adrianto, L., Kurniawan, F., Romadhon, A., Bengen, D. G., Sjafrie, N. D. M., Damar, A., & Kleinertz, S. (2021). Assessing social-ecological system carrying capacity for urban small island tourism: The case of Tidung Islands, Jakarta Capital Province, Indonesia. *Ocean & Coastal Management, 212*(October), 105844.

Alkier, R., Milojica, V., & Roblek, V. (2022). The complexity of the tourism supply chain in the 21st century: A bibliometric analysis. *Kybernetes, 52*(11), 5480–5502.

Baddour, J., & Percebois, J. (2012). Insularity and sustainable transport: Challenges and perspectives – Reunion Island, a case in point. *International Journal of Energy Sector Management, 6*(4), 558–568.

Bardolet, E., & Sheldon, P. J. (2008). Tourism in archipelagos: Hawai'i and the Balearics. *Annals of Tourism Research, 35*(4), 900–923.

Barrowclough, D. (2007). Foreign investment in tourism and small island developing states. *Tourism Economics, 13*(4), 615–638.

Becken, S., Whittlesea, E., Loehr, J., & Scott, D. (2020). Tourism and climate change: Evaluating the extent of policy integration. *Journal of Sustainable Tourism, 28*(10), 1603–1624.

Belle, N., & Bramwell, B. (2005). Climate change and small island tourism: Policy maker and industry perspectives in Barbados. *Journal of Travel Research, 44*(1), 32–41.

Bernard, K., & Cook, S. (2015). Luxury tourism investment and flood risk: Case study on unsustainable development in Denarau island resort in Fiji. *International Journal of Disaster Risk Reduction, 14*(December), 302–311.

Bernardo, E., & Jorge, F. (2019). Are local residents able to contribute to tourism governance? – Impacts and perceptions in Cape Verde. *PASOS Revista de Turismo y Patrimonio Cultural, 17*(3), 611–624.

Bohlin, M., Brandt, D., & Elbe, J. (2014). The development of Swedish tourism public policy 1930–2010. *Scandinavian Journal of Public Administration, 18*(1), 19–39.

Boukas, N., & Ziakas, V. (2016). Tourism policy and residents' well-being in Cyprus: Opportunities and challenges for developing an inside-out destination management approach. *Journal of Destination Marketing & Management, 5*(1), 44–54.

Brida, J. G., Cortes-Jimenez, I., & Pulina, M. (2016). Has the tourism-led growth hypothesis been validated? A literature review. *Current Issues in Tourism, 19*(5), 394–430.

Butler, R. W. (1980). The concept of a tourist area cycle of evolution: Implications for management of resources. *Canadian Geographer/Le Géographe canadien, 24*(1), 5–12.

Castellani, V., & Sala, S. (2010). Sustainable performance index for tourism policy development. *Tourism Management, 31*(6), 871–880.

Charles, A., Darné, O., & Hoarau, J.-F. (2019). How resilient is La Réunion in terms of international tourism attractiveness: An assessment from unit root tests with structural breaks from 1981–2015. *Applied Economics, 51*(24), 2639–2653.

Church, A. (2004). Local and regional tourism policy and power. In A. A. Lew, C. M. Hall, & A. M. Williams (Eds.), *A companion to tourism* (pp. 555–568). Malden, MA: Blackwell Publishing.

Dela Santa, E., & Saporsantos, J. (2016). Philippine tourism act of 2009: Tourism policy formulation analysis from Multiple Streams. *Journal of Policy Research in Tourism, Leisure and Events, 8*(1), 53–70.

Dwyer, L. (2015). Computable general equilibrium modelling: An important tool for tourism policy analysis. *Tourism and Hospitality Management, 21*(2), 111–126.

Edgell, D. L. (1987). The formulation of tourism policy-a managerial framework. In J. R. B. Ritchie & C. R. Goeldner (Eds.), *Travel, tourism, and hospitality research. A handbook for managers and researchers* (pp. 23–33). Washington, DC: CABI.

Edgell, D. L. (2022, August 6, 2022). [Ten important World tourism issues for 2022].

Farmaki, A., Altinay, L., & Yaşarata, M. (2016). Rhetoric versus the realities of sustainable tourism: The case of Cyprus. In P. Modica & M. Uysal (Eds.), *Sustainable island tourism: Competitiveness, and quality-of-life* (pp. 35–50). Wallingford, UK: CABI.

Farsari, I. (2012). The development of a conceptual model to support sustainable tourism policy in north Mediterranean destinations. *Journal of Hospitality Marketing & Management, 21*(7), 710–738.

Fauzel, S., Seetanah, B., & Sannassee, R. V. (2017). Analysing the impact of tourism foreign direct investment on economic growth: Evidence from a small island developing state. *Tourism Economics, 23*(5), 1042–1055.

Font, X., Elgammal, I., & Lamond, I. (2017). Greenhushing: The deliberate under communicating of sustainability practices by tourism businesses. *Journal of Sustainable Tourism, 25*(7), 1007–1023.

Garcia, F. A. (2013). Tourism policy and territorial imbalances in Spain. *Bulletin of Geography. Socio-economic Series, 22*(22), 7–19.

Garcia, F. A. (2014). A comparative study of the evolution of tourism policy in Spain and Portugal. *Tourism Management Perspectives, 11*(July), 34–50.

Gill, A. (2004). Tourism communities and growth management. In A. A. Lew, C. M. Hall, & A. M. Williams (Eds.), *A companion to tourism* (pp. 569–583). Malden, MA: Blackwell Publishing.

Grilli, G., Tyllianakis, E., Luisetti, T., Ferrini, S., & Turner, R. K. (2021). Prospective tourist preferences for sustainable tourism development in small island developing states. *Tourism Management, 82*(February), 104178.

Gruetzmacher, I. (2021). *Islands and sustainable tourism policies: A global exploration.* Master of Science. University of Groningen, The Netherlands.

Hall, C. M. (2009). Innovation and tourism policy in Australia and New Zealand: Never the twain shall meet? *Journal of Policy Research in Tourism, Leisure and Events, 1*(1), 2–18.

Hall, C. M. (2011). A typology of governance and its implications for tourism policy analysis. *Journal of Sustainable Tourism, 19*(4–5), 437–457.

Hambira, W. L., & Saarinen, J. (2015). Policy-makers' perceptions of the tourism–climate change nexus: Policy needs and constraints in Botswana. *Development Southern Africa (Sandton, South Africa), 32*(3), 350–362. doi:10.1080/0376835X.2015.1010716.

Hassan, A., & Burns, P. (2014). Tourism policies of Bangladesh—A contextual analysis. *Tourism Planning & Development, 11*(4), 463–466.

Jenkins, C. L. (2015). Tourism policy and planning for developing countries: Some critical issues. *Tourism Recreation Research, 40*(2), 144–156.

Kumar, N. N., & Patel, A. (2023). Nonlinear effect of air travel tourism demand on economic growth in Fiji. *Journal of Air Transport Management, 109*(June), 102402.

Lickorish, L. J. (1991). Developing a single European tourism policy. *Tourism Management, 12*(3), 178–184.

Lim, C. C., & Cooper, C. (2009). Beyond sustainability: Optimising island tourism development. *International Journal of Tourism Research, 11*(1), 89–103.

Mai, T., & Smith, C. (2015). Addressing the threats to tourism sustainability using systems thinking: A case study of Cat Ba Island, Vietnam. *Journal of Sustainable Tourism, 23*(10), 1504–1528.

Mason, P. (2020). *Tourism impacts, planning and management.* Abingdon: Routledge.

McLeod, M. (2021). The Bahamas: Tourism policy within a pandemic. In *COVID in the islands: A comparative perspective on the Caribbean and the Pacific* (pp. 219–230). Singapore: Springer.

McLeod, M. (2023). Tourism policy networks in four Caribbean countries. *Annals of Tourism Research Empirical Insights, 4*(2), 100113.

McLeod, M., Chambers, D., & Airey, D. (2018). A comparative analysis of tourism policy networks. In M. McLeod & R. Croes (Eds.), *Tourism management in warm-water island destinations* (pp. 77–94). Wallingford, UK: CABI.

Nyakunu, E. P. (2014). *The role of small and medium enterprises (SMEs) in tourism policy formulation in Namibia.* University of Johannesburg, South Africa.

Page, S. J. (2014). *Tourism management.* Abingdon, UK: Routledge.

Peterson, R. R. (2020). Over the Caribbean top: Community well-being and over-tourism in small island tourism economies. *International Journal of Community Well-Being, 6*(November), 1–38.

Petit Futé (2023). Discover réunion: Current issues. Retrieved from www.petitfute.co.uk/p9-reunion/decouvrir/d1194-current-issues/

Pratt, S., Gibson, D., & Movono, A. (2013). Tribal tourism in Fiji: An application and extension of Smith's 4Hs of indigenous tourism. *Asia Pacific Journal of Tourism Research, 18*(8), 894–912.

Richards, G. (1995). Politics of national tourism policy in Britain. *Leisure Studies, 14*(3), 153–173. doi:10.1080/02614369500390131

Schubert, S. F., Brida, J. G., & Risso, W. A. (2011). The impacts of international tourism demand on economic growth of small economies dependent on tourism. *Tourism Management, 32*(2), 377–385.

Stevenson, N., Airey, D., & Miller, G. (2009). Complexity theory and tourism policy research. *International Journal of Tourism Policy, 2*(3), 206–220.

Tandrayen-Ragoobur, V., & Fauzel, S. (2021). Climate change, governance and economic growth: the case of small island developing states. *Small States & Territories, 4*(2), 245–258. Retrieved from www.um.edu.mt/library/oar/handle/123456789/83374

Telesford, J. N. (2022). Restructuring island tourism: Using the socioeconomic metabolism (SEM) and multilevel perspective (MLP) as models for transitioning to sustainable island tourism. In I. Bethell-Bennett, S. Rolle, J. Minnis, & F. Okumus (Eds.), *Pandemics, disasters, sustainability, tourism* (pp. 109–123). Leeds, UK: Emerald Publishing.

United Nations (2013). Réunion Island declaration on sustainable tourism in islands. In *Cotswolds sustainable tourism workshops* (p. 5). New York: United Nations.

Walker, T. B., Lee, T. J., & Li, X. (2021). Sustainable development for small island tourism: Developing slow tourism in the Caribbean. *Journal of Travel & Tourism Marketing, 38*(1), 1–15.

Wisnumurti, A. A. G. O., Darma, K., & Putra, I. N. G. M. (2021). Tourism policy and the impact of tourism on Bali Island. *Journal of Hunan University Natural Sciences, 47*(12), 95–104.

Xing, Y., & Dangerfield, B. (2018). Modelling the sustainability of mass tourism in island tourist economies. In M. Kunc (Ed.), *System dynamics: Soft and hard operational research* (pp. 303–327). London: Springer.

Zhang, H. Q., Chong, K., & Ap, J. (1999). An analysis of tourism policy development in modern China. *Tourism Management, 20*(4), 471–485.

5   Tourism policy capacity building

5.1 Introduction

This chapter considers the required capacities for tourism policies within island destinations. Policies are needed to warden off sustainability threats (Roxas, Rivera, & Gutierrez, 2020). Building capacity for policies involves understanding the policy processes. A policy process is an activity to deliver effective policies to realize results. Pathak, van Beynen, Akiwumi, and Lindeman (2022) have noted that in the case of The Bahamas, policy innovation and implementation have been constrained by funding and human capacity. The capacity to deliver policies is related to the policy content as well as the interactions of tourism stakeholders. Within sustainable tourism governance, planning, networking, and capacity building are important activities (Dwyer, 2018).

Four policy processes exist: policy-making, policy formulation, policy implementation, and policy taking (McLeod, 2023), and each process must be detailed to reveal the needed capabilities to improve the process. Policy-making is perhaps the most popular activity amongst the policy processes. Policymakers determine and select policies that are taken on board to bring about some desired change or result. Policymakers are usually part of the public sector management machinery, and policymakers may also belong to other organizations such as development banks. Concerns about power relations and ethical approaches to policy-making processes have been raised (Liasidou, 2019). Policy formulators are the researchers in the policy development process. Policy formulators include academics, researchers, professionals, and other specialists with a wide range of expertise to inform policy. Policy formulators conduct research, consultation, and engage stakeholders who will directly be affected by the formulated policies. Building capacity for policy formulation through multi-stakeholder partnerships between government, private sector, and non-governmental organizations is important for realization of the Sustainable Development Goals (SDGs) in Fiji (Movono & Hughes, 2022).

Policy implementation is the action of putting policies into force. The ways in which policies can be implemented include laws, regulations, incentives, and so on. The goal of policy implementation is to bring about a change that will achieve policy goals. Policy implementation may be constrained by bureaucratic processes, an unwillingness to participate, and a lack of monitoring of the implementation

DOI: 10.4324/9781003435112-6

process. Do and Phi (2022) have noted ambiguity and multiple responsibility in implementing island tourism policy. The policy process that is often not considered is policy taking. Policy taking may also exist when stakeholders are not engaged with other policy processes, making, formulation, and implementation (McLeod, 2023). The policies are therefore adopted without any input from a policy taker. Tosun (2001) has taken the view that in sustainable tourism development tourism stakeholders are decision-makers rather than decision-takers.

5.2 Island tourism policy-making

Policy-making must be distinguished from the other policy processes as an activity that involves being accountable for policy decisions. A policymaker has a specific role in policy development, and policy-making should be studied in relation to island tourism. In developing countries policy-making is influenced by policy elites, issue formation, selection criteria, and policy circumstances (Grindle & Thomas, 1989). Policymakers may serve on a state board with overall responsibility for tourism in the country. Capacity building for policy-making needs to be addressed in the tourism literature to improve policy processes. A range of issues and challenges can stymie tourism development. Policies must be timely and require analysis to determine appropriate policies to guide tourism growth and development. Elliott and Neirotti (2008) have referred to the indecisive nature of policy formulation. Timeliness and relevance of tourism policies to facilitate tourism development are essential. Some of the key questions include

- Are the tourism policies appropriate to address the major tourism issues and challenges of tourism growth and development in islands? This relates to the contents of tourism policies in island tourism environments.
- Are the existing tourism policy actors able to influence policies? This relates to the connections amongst the policy actors in islands.
- Are other actors needed for effective tourism policies? This relates to the composition of tourism actors involved with policy-making.

The tourism literature has addressed the issue of policy-making by looking into the policymakers, reviewing the mechanisms of the policy-making process, considering policy implementation, and so on. Whilst these perspectives can contribute to improving the effectiveness of tourism policies, there has been limited attention given to building capacity for appropriate policies for island tourism development. Kennedy, Crawford, Main, Gauci, and Schembri (2022) have noted the importance of stakeholder hazard awareness and vulnerability assessment in small island destinations such as Malta to develop appropriate strategies for implementation. Capacity building in this context means the resources, such as knowledge, technology, training, and so on, are provided to policymakers. Capacity building is important for environments that are resource constrained, including limited financial and human capital for development. Development means the expansion of existing activities that would benefit the economy (Buhalis, 1999).

In the process of tourism development there are different courses of action. Know how about an appropriate path should not be learnt, but should be built by design, and herein is the value of capacity building for policy-making. First, a review of the existing capacity to develop policies in an island tourism destination must be conducted. For this, the range of tourism operators and the structure of the tourism industry should be considered, and existing and proposed policies inventoried. Second, the vision and mission in terms of where the tourism industry should be in 5, 10, or 20 years should be articulated. In the context of sustainable tourism, Sharpley (2003) has suggested that small island states should consider mass tourism as an appropriate development path to optimize the contribution of tourism. A vision and a mission support identification of any policy gaps. As policy gaps emerge from an evaluative process, identification and prioritization of the policy issues support achievement of the stated vision and mission. Third, review existing and alternative policies that could address the policy gaps. This can be done by a review of a body of policies, the experience of implementing these, and likely results. A particular policy direction should be identified and clarified.

Fourth, the necessary governance structures to implement the policies are analysed. These can include the formal institutional structures and informal processes that occur in the island destination during policy implementation. Fifth, review the legal and regulatory framework and other facilitators that will assist the policy-making process including the implementation of tourism policies should be done. Without understanding a facilitating framework, policy-making can be delayed, or stop having any effect. Sixth, policy-making resources are identified and deployed to build capacity. Resource allocation should be viewed as a partnership among the various stakeholders and not the sole responsibility of the public sector. The selection of actors and the required mechanisms to support effective policy-making should be addressed. With appropriate allocation of resources, a balance can be obtained, and determination of incapacities made. Wong, Jiang, Klint, Dominey-Howes, and DeLacy (2013) have noted that successful climate adaptation policy-making, identification of actors and institutions are needed to support the policy mechanisms. Thereby, a course for mediating and building new capacities can be taken. Seventh, areas that need to be addressed to improve the policy-making capacity will be actioned. Improvement activities involve actors and institutions that create policies for the sustainable development of island destinations.

5.3 Island tourism policy formulation

Within the tourism policy formulation arena, there are competing interests (Church, Ball, Bull, & Tyler, 2000), and understanding the interactions among policy actors become important to improve the policy formulation process. Church et al. (2000) have pointed to non-integration of policy actions because of 'institutional thickness'. Institutions in the tourism sector should become flexible, engaging with stakeholder interests in the policy formulation process and create networking opportunities for tourism actors to engage with policy discussions. Tourism policies are formed when actors relate and interrelate. Tourism policy formulation activities occur within policy networks through collaborations and partnerships between public, private, not-for-profit agents, and other stakeholders (Elliott, 2020; McLeod, 2023). Connections among the multitude of businesses who are in direct contact with tourists should be integrated into tourism policy formulation activities. A lack of well-formulated tourism policies affects tourism development (McLeod & Airey, 2007). Collaboration between policy formulators is an important element in the tourism sector's growth and sustainability.

Formulated policies are managed in various ways including through tourism planning instruments, policy documents, and governance institutions. The Government of the Commonwealth of Dominica have articulated a National Tourism Policy and Tourism Master Plan (Commonwealth of Dominica, 2013a, 2013b). A Tourism Master Plan for Jamaica states, "The Plan seeks to guide the industry's development over the next decade by creating a strategic vision for its growth and development and establishing an enabling environment to help it realise that vision" (Commonwealth Secretariat, 2002, p. iii). The Republic of Maldives has articulated the Maldives Fifth Tourism Master Plan 2023–2027: Goals and Strategies and has stated 15 priority goals with a stated vision of "Maldives strives to lead the world in sustainable tourism" (Government of the Republic of Maldives, 2023, p. 10). Island tourism development is supported by a range of policies, nonetheless, Joppe (2018) has noted the underutilization of Organization for Economic Cooperation and Development (OECD) database about tourism policies. Tourism policy formulation is directly related to tourism governance activities. Policies are formulated in relation to responsibilities of stewardship and tourism planning and development activities. Tourism governance involves both internal and external entities that are concerned with a particular tourist destination (Beaumont & Dredge, 2010). Governance of tourism activities affects sustainability (Bramwell & Lane, 2013). Hall (2011) has framed tourism governance types that are applicable and has suggested two steering modes to govern tourism: hierarchical (hierarchies and markets) or non-hierarchical (networks and communities). Stakeholder involvement in tourism policy formulation and implementation is not a simple matter, since a high level of complexity makes the processes difficult to understand (Do & Phi, 2022). As such, considerations about the institutions involved in managing island tourism policy formulation are important since policy formulation involves a process (Figure 5.1).

The structure of government is important for tourism policy formulation (Do & Phi, 2022). The state has a role in tourism policy formulation to correct market failures, provide public goods, and avoid information asymmetry (Socher, 2001). The role of the public sector in tourism governance cannot be over-estimated, but what is unclear is whether that governance should be centralized or decentralized (Yüksel, Bramwell, & Yüksel, 2005). It has been suggested that decentralization results in better decision-making, but then there are issues of control and power (Yüksel et al., 2005). The impact of climate change susceptibility is an important

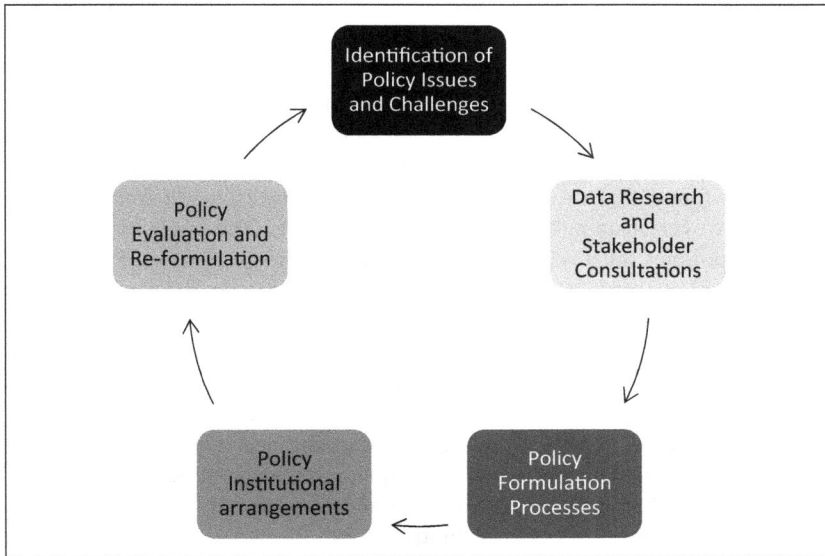

Figure 5.1 A tourism policy formulation process.

area for government to formulate policies (Moore, 2010). Sustainable develop-ment policies for island destinations must consider vulnerability to changes in temperature, rising sea levels, and effects on marine life and coral reefs (Moreno & Becken, 2009). Power shifts can divert policy formulation activities between institutions and resource management and thus affect sustainability (Shakeela, 2022; Silva, 2015).

5.4 Island tourism policy implementation

Policy implementation involves actioning stated policies, and stakeholder buy-in is essential for effective implementation. Policy implementation is fundamentally pol-itical and involves party politics and crony capitalism (Muangasame & McKercher, 2015). In such circumstances, the identification of suitable agencies to execute the policies is very important for effective implementation. One of the criticisms of public policy implementation is the bureaucracy involved in getting the policy actions that are needed to maximize the benefits and minimize the costs of tourism development. An example of this is the bureaucracy involved in accessing tax and other benefits under tourism development legislation. Singgalen, Sasongko, and Wiloso (2018) have found that in the remote area of North Halmahera, the bur-eaucratic system, human resources, funding, and coordination were important for policy implementation. The requirements could be arduous and restrictive, and so an incentive acts as a disincentive because stakeholders may not have the resource capacity to meet all requirements. Some countries set up investment agencies that facilitate a one-stop shop to assist investors in navigating the bureaucratic processes.

Another challenge is the actual interpretation of a policy instrument by those who must act and implement the policy. Pickel-Chevalier, Bendesa, and Putra (2022) have reviewed an integrated tourist village model policy for Bali, and after 25 years found that three villages have been successful in implementing the policy, however sustainability remains a challenge. The capacity to implement tourism policies must be supported by building the awareness of the public and training implementers in understanding the policy requirements. Challenges of policy implementation include the capital and labour resources that set up the facilitating infrastructure for policy implementation. Financial resources to apply technological solutions for policy implementation is one consideration, and funding is a significant concern for implementation of climate policies (Pathak et al., 2022; Wolf et al., 2022). Additionally, mapping and drawing a flow chart that can guide tourism stakeholders in the implementation process can be useful. As such, materials that explain the rationale and intended outcomes of tourism policies are needed to assist with facilitating policy implementation. Additional challenges of policy implementation include the simplicity of the policy solutions. A clear understanding of the policy content and how it addresses the problem can assist with its implementation.

Political party changes affect policy implementation, and external forces that affect the economic, social, or environmental circumstances in island destinations can also affect policy implementation (Chambers & Airey, 2001; McLeod & Airey, 2007; Movono & Scheyvens, 2022; Velasco, 2016). For example, a policy relating to creating greater linkages to build an inclusive destination can be affected by rising costs. An effective monitoring mechanism is needed for policy implementation to ensure the barriers to smooth implementation are identified. Twining-Ward and Butler (2002) have suggested designing flexibility in monitoring and implementation. Policies may require adjustment or a total policy change. In addition, policy implementation should be supported by monitoring the results of the policies to ensure those tourism stakeholders are receiving the intended benefits from the policies. In the Shetland Islands, Leask and Rihova (2010) have suggested that policy implementation has to be supported by precise timelines, guidelines, consultation, and regular monitoring. Data collected on a timely basis and in a continuous manner help with understanding the effectiveness of the tourism policies. For example, if a policy is instituted to develop certain types of accommodation in the destination, then a monitoring of the new developments will assist with an assessment of the policy's effectiveness. Regarding policy implementation, the environmental, social, and governance (ESG) framework provides policy guidance in evaluating contributions to sustainability. The mini case study explores the ESG framework.

Mini Case Study 4: Adoption of environmental, social, and governance (ESG) framework

An Island Innovation Forum discussed local legislation and external global regulations for ESG framework (Island Innovation, 2023). An ESG framework is the concept that businesses will focus on the environmental, social, and governance aspects in their operations and thereby contribute to balancing sustainability (Organization for Economic Cooperation and Development, 2017). An ESG framework shifts focus from profitability towards society receiving a greater benefit. Ternel and Greyling (2018) have argued that small island developing states still face issues about sustainability that warrant consideration of policies that contribute to poverty alleviation and social inclusion. Addressing matters relating to environmental, society well-being, and governance can guide achievement of poverty alleviation and social inclusion (Dossou, Ndomandji Kambaye, Bekun, & Eoulam, 2023). Nonetheless, although tourism has a substantial ecological or social impact (Kapoor & Wangdus, 2022), the ESG concept's application to the tourism industry has been limited. Resources can be found in the World Bank's data bank about ESG data (World Bank, 2023).

Mini Case Study 4 discussion questions

1 Using data from the World Bank dataset, create an ESG framework for an island tourism destination.
2 Identify and categorize the key policies that are needed to implement ESG in an island destination.
3 Evaluate an institutional framework that would be needed to formulate and implement ESG policies in island destinations.

5.5 Island tourism policy taking

Policy taking can materialize in changing circumstances of market conditions (Pastras & Bramwell, 2013). An example of this is an international tour operator contributing to policies in the destination based on changing circumstances in the tourist generating market. Policy taking is also local stakeholders not contributing to the policy decisions being formulated and implemented in an island destination (McLeod, 2023). Policy taking may also involve activities that local stakeholders may not be involved with. On one hand, foreign direct investment policies in small island developing states can realize several impacts of consumer demand, capital formation, human resources, and procurement linkages (Barrowclough, 2007). On the other hand, the remoteness and small domestic economies of small island states may make foreign investment risky (Bojanic &

Lo, 2016). In addition, the extent to which stakeholders in the domestic economy agree with such a policy may result in policy conflict. As a result, policy taking may also occur when there are limits to local community involvement in tourism decision-making (Tosun, 2000).

An understanding of the power of policy actors provides a basis for determining whether a stakeholder policy takes (McLeod, 2023). A network approach allows for a determination of the more powerful actors that influence tourism development. Policy taking may be viewed as a matter of power imbalances. The power an entity has over another is based on dependence of one on another (Emerson, 1962). Those policy actors who are more dependent on others for resources may have less of a voice in the other policy processes. The policy taking process is very important as it connects all actors within the tourism sector in a working relationship to head towards a sustainable development path. If the policy taking activities are not in sync with formulation, implementation, and making, the tourism industry may not achieve sustainable development. Building capacity for policy taking involves building consensus among tourism industry stakeholders and coordinating awareness of the benefit of policies. A position of policy taking may also disadvantage the tourism industry's progress as stakeholders' inputs in tourism actions may be limited. Stakeholders need platforms to contribute to decision-making in the tourism sector (Tosun, 2000). Policy taking may also replicate the social and cultural inequalities in island countries.

5.6 Conclusion

Four policy processes are involved in island destinations including policy-making, formulation, implementation, and policy taking. For realizing sustainable tourism development, island destinations must build capacities to improve the four policy processes. Tourism policies that seek to support sustainable development must be well-formulated to be effective and receive widespread support to be adopted. For example, adoption of the SDGs in island tourism, and realization of the benefits, means that policies in the different areas of the SDGs must be addressed by building the four policy processes. Capacity building assists with the coordination and integration of tourism policies by ensuring that plans and policies are workable in achievement of the island destination's goals.

Chapter 5 discussion questions

1 Explain the role of policymakers in an island destination.
2 Apply the policy formulation process framework to achieve sustainable development in an island destination.
3 Discuss the capacity barriers in the effective implementation of tourism policies in an island destination.

References

Barrowclough, D. (2007). Foreign investment in tourism and small island developing states. *Tourism Economics, 13*(4), 615–638.

Beaumont, N., & Dredge, D. (2010). Local tourism governance: A comparison of three network approaches. *Journal of Sustainable Tourism, 18*(1), 7–28.

Bojanic, D. C., & Lo, M. (2016). A comparison of the moderating effect of tourism reliance on the economic development for islands and other countries. *Tourism Management, 53*(April), 207–214.

Bramwell, B., & Lane, B. (2013). *Tourism governance: Critical perspectives on governance and sustainability*. Abingdon: Routledge.

Buhalis, D. (1999). Tourism on the Greek Islands: Issues of peripherality, competitiveness and development. *International Journal of Tourism Research, 1*(5), 341–358.

Chambers, D., & Airey, D. (2001). Tourism policy in Jamaica: A tale of two governments. *Current Issues in Tourism, 4*(2–4), 94–120.

Church, A., Ball, R., Bull, C., & Tyler, D. (2000). Public policy engagement with British tourism: The national, local and the European Union. *Tourism Geographies, 2*(3), 312–336.

Commonwealth of Dominica (2013a). *National tourism policy 2020*. Dominica: Government of the Commonwealth of Dominica. Retrieved from https://tourism.gov.dm/images/documents/national_tourism_policy_june_2013.pdf

Commonwealth of Dominica (2013b). *Tourism master plan 212–2022*. Dominica: Government of the Commonwealth of Dominica. Retrieved from https://tourism.gov.dm/images/documents/tourism_master_plan/tourism_master_plan_june2013.pdf

Commonwealth Secretariat (2002). *Master plan for sustainable tourism development*. UK: Commonwealth Secretariat. Retrieved from www.tpdco.org/wp-content/uploads/2016/07/Master-Plan-for-Sustainable-Tourism-Development.pdf

Do, H., & Phi, G. T. (2022). Marine and island tourism "Stakeholder involvement in policy formulation and implementation". In H. T. Bui, G. T. Phi, L. H. Pham, H. H. Do, A. Le, & B. Nghiem-Phu (Eds.), *Vietnam tourism: Policies and practice* (pp. 63–84). Wallingford, UK: CABI.

Dossou, T. A. M., Ndomandji Kambaye, E., Bekun, F. V., & Eoulam, A. O. (2023). Exploring the linkage between tourism, governance quality, and poverty reduction in Latin America. *Tourism Economics, 29*(1), 210–234.

Dwyer, L. (2018). Emerging ocean industries: Implications for sustainable tourism development. *Tourism in Marine Environments, 13*(1), 25–40.

Elliott, J. (2020). *Tourism: Politics and public sector management*. Abingdon, UK: Routledge.

Elliott, S. M., & Neirotti, L. D. (2008). Challenges of tourism in a dynamic island destination: The case of Cuba. *Tourism Geographies, 10*(3), 375–402.

Emerson, R. M. (1962). Power-dependence relations. *American Sociological Review, 27*(1), 31–41.

Government of the Republic of Maldives (2023). Maldives fifth tourism master plan 2023–2027: Goals and strategies. Republic of Maldives. Retrieved from www.tourism.gov.mv/dms/document/4969b4831928f1bdf3506340fb6974fc.pdf

Grindle, M. S., & Thomas, J. W. (1989). Policy makers, policy choices, and policy outcomes: The political economy of reform in developing countries. *Policy Sciences, 22*(3–4), 213–248.

Hall, C. M. (2011). A typology of governance and its implications for tourism policy analysis. *Journal of Sustainable Tourism, 19*(4–5), 437–457.

Island Innovation (2023, April 19, 2023). The evolution of ESG: The future of sustainable development policy. Retrieved from https://islandinnovation.co/videos/the-evolution-of-esg-the-future-of-sustainable-development-policy/

Joppe, M. (2018). Tourism policy and governance: Quo vadis? *Tourism Management Perspectives, 25*(January), 201–204.

Kapoor, S., & Wangdus, J. (2022). Impact of overtourism on sustainable development and local community wellbeing in the Himalayan region. *International Journal of Leisure and Tourism Marketing, 7*(3), 199–214.

Kennedy, V., Crawford, K. R., Main, G., Gauci, R., & Schembri, J. A. (2022). Stakeholder's (natural) hazard awareness and vulnerability of small island tourism destinations: A case study of Malta. *Tourism Recreation Research, 47*(2), 160–176.

Leask, A., & Rihova, I. (2010). The role of heritage tourism in the Shetland Islands. *International Journal of Culture, Tourism and Hospitality Research, 4*(2), 118–129.

Liasidou, S. (2019). Understanding tourism policy development: A documentary analysis. *Journal of Policy Research in Tourism, Leisure and Events, 11*(1), 70–93.

McLeod, M. (2023). Tourism policy networks in four Caribbean countries. *Annals of Tourism Research Empirical Insights, 4*(2), 100113.

McLeod, M., & Airey, D. (2007). The politics of tourism development: A case of dual governance in Tobago. *International Journal of Tourism Policy, 1*(3), 217–231.

Moore, W. R. (2010). The impact of climate change on Caribbean tourism demand. *Current Issues in Tourism, 13*(5), 495–505.

Moreno, A., & Becken, S. (2009). A climate change vulnerability assessment methodology for coastal tourism. *Journal of Sustainable Tourism, 17*(4), 473–488.

Movono, A., & Hughes, E. (2022). Tourism partnerships: Localizing the SDG agenda in Fiji. *Journal of Sustainable Tourism, 30*(10), 2318–2332.

Movono, A., & Scheyvens, R. (2022). Tourism and politics: Responses to crises in Island states. *Tourism Planning & Development, 19*(1), 50–60.

Muangasame, K., & McKercher, B. (2015). The challenge of implementing sustainable tourism policy: A 360-degree assessment of Thailand's "7 Greens sustainable tourism policy". *Journal of Sustainable Tourism, 23*(4), 497–516.

Organization for Economic Cooperation and Development (2017). Investment governance and the integration of environmental, social and governance factors. Retrieved from www.oecd.org/finance/Investment-Governance-Integration-ESG-Factors.pdf

Pastras, P., & Bramwell, B. (2013). A strategic-relational approach to tourism policy. *Annals of Tourism Research, 43*(October), 390–414.

Pathak, A., van Beynen, P. E., Akiwumi, F. A., & Lindeman, K. C. (2022). Climate change in the strategic tourism planning for small islands: Key policy actors' perspectives from The Bahamas. In I. Bethell-Bennett, S. Rolle, J. Minnis, & F. Okumus (Eds.), *Pandemics, Disasters, Sustainability, Tourism* (pp. 125–143). Leeds, UK: Emerald Publishing.

Pickel-Chevalier, S., Bendesa, I. K. G., & Putra, I. N. D. (2022). The integrated touristic villages: An Indonesian model of sustainable tourism? In M. McLeod, R. Dodds, & R. Butler (Eds.), *Island tourism sustainability and resiliency* (pp. 262–286). Abingdon: Routledge.

Roxas, F. M. Y., Rivera, J. P. R., & Gutierrez, E. L. M. (2020). Framework for creating sustainable tourism using systems thinking. *Current Issues in Tourism, 23*(3), 280–296.

Shakeela, A. (2022). *Policies addressing climate change and sustainable tourism outcomes: Case of the Maldives.* Paper presented at the CAUTHE 2022 Conference Online: Shaping the Next Normal in Tourism, Hospitality and Events: Handbook of Abstracts of the 32nd Annual Conference, Online.

Sharpley, R. (2003). Tourism, modernisation and development on the island of Cyprus: Challenges and policy responses. *Journal of Sustainable Tourism, 11*(2–3), 246–265.

Silva, L. (2015). How ecotourism works at the community-level: The case of whale-watching in the Azores. *Current Issues in Tourism, 18*(3), 196–211.

Singgalen, Y. A., Sasongko, G., & Wiloso, P. G. (2018). Tourism destination in remote area: Problems and challenges of tourism development in North Halmahera as remote and border areas of Indonesia–Philippines. *Journal of Indonesian Tourism and Development Studies, 6*(3), 175–186.

Socher, K. (2001). What are the tasks of the state in providing the framework for tourism? *Tourism Review, 56*(1/2), 57–60.

Ternel, M., & Greyling, L. (2018). *An assessment of sustainable tourism and its opportunities in Mauritius.* Paper presented at the International Conference on Tourism Research, Kavala, Greece.

Tosun, C. (2000). Limits to community participation in the tourism development process in developing countries. *Tourism Management, 21*(6), 613–633.

Tosun, C. (2001). Challenges of sustainable tourism development in the developing world: The case of Turkey. *Tourism Management, 22*(3), 289–303.

Twining-Ward, L., & Butler, R. (2002). Implementing STD on a small island: Development and use of sustainable tourism development indicators in Samoa. *Journal of Sustainable Tourism, 10*(5), 363–387.

Velasco, M. (2016). Dynamics of Spanish tourism policy: The political system as a driver and policy instruments as indicators of change (1952–2015). *International Journal of Tourism Policy, 6*(3–4), 256–272.

Wolf, F., Moncada, S., Surroop, D., Shah, K. U., Raghoo, P., Scherle, N., … Havea, P. H. (2022). Small island developing states, tourism and climate change. *Journal of Sustainable Tourism*, 1–19. doi:10.1080/09669582.2022.2112203.

Wong, E., Jiang, M., Klint, L. M., Dominey-Howes, D., & DeLacy, T. (2013). Evaluation of policy environment for climate change adaptation in tourism. *Tourism and Hospitality Research, 13*(4), 201–225.

World Bank (2023). Environment social and governance (ESG) data. Retrieved from https://databank.worldbank.org/source/environment-social-and-governance-(esg)-data

Yüksel, F., Bramwell, B., & Yüksel, A. (2005). Centralized and decentralized tourism governance in Turkey. *Annals of Tourism Research, 32*(4), 859–886.

Part II
Island tourism governance

6   Tourism governance structures and stakeholders

6.1 Introduction

The first part of this book explored island tourism policy through perspectives of demand, supply, sustainability policy, and policy capacity building. This second part considers the stakeholders of island tourism governance. Governance is an important aspect of sustainable development as decisions are being made and actions are being taken to sustain island tourism. Douglas (2006) has noted two governance assumptions of sustainable development that affect small island states: the first assumption is that island societies are homogenous and this likely results in compliance with resource management; the second assumption is that the time horizon of securing resources for future generations suggests the reduction and controlled use of resources (Douglas, 2006). Islands are captive tourism destinations, and governments seek sustainable development in a resource-constrained space. Resource constraints mean that appropriate governance structures are needed to develop and manage a dynamic tourism industry. Uysal and Modica (2016) have pointed out the globalized world and a highly competitive travel and tourism market are key challenges for island destinations. Scheyvens and Momsen (2008) have suggested that comments about small island development focuses on constraints such as a lack of natural resources and a small size and not the strengths of these islands. Key strengths of small island states include beautiful and enticing tourism offerings, good economic performance, high levels of culture, holistic approaches to development, political strength, respect for traditional, social, and natural capital, and strong international linkages (Scheyvens & Momsen, 2008). Stakeholder structures in island tourism governance are explored to harness key strengths of islands.

A range of stakeholders are involved in an island's tourism sector. Partelow and Nelson (2020) have taken the view that island tourism governance is comprised of self-organized social networks to meet the challenges of sustainable development. Based on evolutionary governance theory and within the context of challenges of tourism growth on islands, the role of multi-level governance and institutional change can be understood (Partelow & Nelson, 2020). Governance structures are evolving in island tourism sustainable development. First, a governance framework is based on the organization of government. Kerr (2005) has presented an island

DOI: 10.4324/9781003435112-8

autonomy framework from fully incorporated, generally smallest islands (e.g. Lindisfarne, Orkney Islands, and Galapagos) to fully independent states, generally largest island (e.g. Cyprus, Samoa, and New Zealand) with increasing political and economic autonomy. Most if not all islands were colonized, and this resulted in the setting up of government arrangements linked to the mainland. In the case of the Caribbean and South Pacific, islands have been governed by Dutch, English, French, German, or Spanish governments. Second, governance arrangements emerge from a political economy of relationships among the various tourism actors making decisions that relate to how the tourism industry operates and functions (Valdivielso & Moranta, 2019). In this regard, island governance arrangements may reflect the structures of the dominant market entities involved in international tourism. In island tourism, a suggestion of decolonization and reimagined tourism means new governance structures (Higgins-Desbiolles, 2022). Foreign domination of an international tourism industry results in limited control of tourism activities in host destinations (Britton, 1982). Governance structures require elaboration to achieve sustainable development in island states.

Island governance structures may take various forms. A tourism industry stakeholders' classification includes hierarchies, markets, networks, and communities (Hall, 2011). Using this classification, this chapter sets out the involvement of a range of tourism industry governance actors in island tourism development and management. Both small and large islands may have nuisances that warrant understanding of the most appropriate governance structure for islands. Becken and Loehr (2022) have pointed out appropriate and suitable governance arrangements for climate action. Synthesizing interview data from Thailand, Pacific Islands, and Mekong, a framework is proposed that outlines three states of governance: (1) supra-state and state governance; (2) network and market governance; and (3) state, network/community, and market governance (Becken & Loehr, 2022). McLeod and Airey (2007) have explored dual governance in a twin-island context and argued for a divergence of policy arrangements on each island. Without exploring the specific dynamics of stakeholder structures, the scope of this chapter is to understand the formal and informal island tourism governance structures that exist, and the effect of these structures on the governance of the tourism sector on islands. An actor-focused approach was taken to explore island tourism governance rather than a spatial governance structure such as local, regional, or national level of governance. Thereby, an actor-led tourism governance path may be created for sustainable development of island tourism destinations.

6.2 Island tourism destination organization

Tourism destination organization studies thematic areas include digitalization, experience, governance, image, loyalty, marketing, and resources (Ivanka, Marion, Anthony, & Rob, 2023). Government entities are the main managers of a tourism destination with responsibility for marketing and product development. Marketing a tourism destination is an expensive venture and such expenditure must be justified based on the level of visitor arrivals. In small islands with limited resources and reduced sources of funding, visitors to the destination may be lower than other islands with greater funding. Graci (2008) using the case of Gili Trawangan, Indonesia, has noted the barriers of sustainable tourism development including, business-owners' lack of momentum, government bureaucracy and corruption, inadequate funds and information, infrastructure impediments, and island culture and isolation of sustainability issues. Successful island destination organization requires strong cooperation and collaboration from stakeholders (Said & Farid, 2022). McLeod (2022) has found that collaboration is essential for island tourism destination recovery after a natural disaster. Destination organization creates the enabling environment for tourism governance that contributes to sustainable development (Becken & Loehr, 2022).

Island tourism destination organizations have a hierarchical structure with a tourism authority or board assigned to govern tourism activities on the island (Table 6.1). A hierarchical structure is evident in handling a crisis (Wan, Li, Lau, & Dioko, 2022). Farsari (2021) has noted that tourism governance requires less governmental control and a multi-stakeholder approach, and has argued that collaborative approaches in governance adopt weak approaches in sustainability with an emphasis on the industry rather than local community. Nonetheless, tourism decision-making bodies contribute to tourism policy and strategies to guide

Table 6.1 Island Destination Organizations (Some Examples)

Island territories/ country	Public sector	Private sector	Non-government sector
Bahamas	Ministry of Tourism, Investments and Aviation	Bahamas Hotel and Tourism Association	Organization for Responsible Governance
Balearic Islands	The Agency for Tourism Strategy of the Balearic Islands	Hotel Business Federation of Mallorca	Marilles Foundation
Fiji	Ministry of Commerce, Trade, Tourism and Transport	Fiji Hotel and Tourism Association	NatureFiji
Galápagos Islands	Governing Council of the Special Regime of Galápagos	International Galápagos Tour Operators Association	Galápagos Conservation Trust
Seychelles	Tourism Seychelles	Seychelles Hospitality and Tourism Association	Seychelles Islands Foundation

Sources: Author's compilation from online sources – The Bahamas Ministry of Tourism (2024), Bahamas Hotel & Tourism Association (2024), Organization for Responsible Governance (2023), Mallorca Sustainable Tourism Observatory (2024a, 2024b), Marilles Foundation (2024), Ministry of Trade Co-operatives Small and Medium Enterprises (2024), Fiji Hotel and Tourism Association (2024), NatureFiji (2024), Government of the Republic of Ecuador (2024), International Galapagos Tour Operators Association (2024), Galápagos Conservation Trust (2024), Tourism Seychelles (2024), Seychelles Hospitality and Tourism Association (2024), Seychelles Islands Foundation (2024).

sustainable development. The composition of the tourism bodies such as boards should be representative of a mix of public and private sector officials. Island tourism governance and destination organization are complicated by sub- and supra-national political bodies of organizations in the case of the Balearic Islands (Valdivielso & Moranta, 2019). Depending on the size of the island and its population, sub-groups or units may be set up to manage the tourism resources in a specific area, with the sub-group or unit being required to report to the main board or authority. In addition, Baldacchino (2004) has pointed out that although small, some islands may be autonomous but not sovereign and therefore function as a sub-national jurisdiction. Sub-national island jurisdictions have specific challenges for tourism development. Amoamo (2013) using a case example of Pitcairn Island has posited the concept of 'decolonising without disengaging' as an empowerment strategy to gain autonomy. Governance of island tourism requires an appropriate political context for workable solutions for sustainable tourism development.

Management and development functions are separated in island destination organization. The overall tourism development of an island may be spread across government ministries with responsibility for planning, infrastructure, transport, and the environment. Good communication and information flow among ministries of a government are essential for climate action for sustainable tourism (Becken & Loehr, 2022). In the case of Cyprus, a national tourism strategy required involvement of ministries with responsibility for industry, commerce, tourism, and department with responsibility for town planning and housing (Liasidou, 2019). As islands are surrounded by water, the role of a ministry with responsibility for aviation becomes of primary importance. Development of a strategic aviation plan for islands in archipelagos cannot be overestimated as islands remotely placed from the capital city may be underserved without proper planning for archipelagic airlift (The Bahamas Department of Aviation, 2023). In addition, climate change adaptation policy relating to aviation is an important consideration within the tourism sector, and government ministries, which are not directly responsible for tourism, complicate the policy-making environment (Klint et al., 2012). An overall island tourism development strategy must be organized and supported by the human resource capital. Typically a national tourism organization model includes a department with responsibility for training and education (Alipour & Kilic, 2005). Development and innovation of tourism products and services require a trained and skilled labour force. The availability of education and training opportunities affect the capacity of tourism supply. The island destination organization must ensure that education and training opportunities are made available at all levels.

6.3 Island tourism markets

Policies to encourage investment and support small businesses are essential to develop thriving markets in island tourism. Peterson (2020) has noted that the expansion of tourism in the Caribbean has been facilitated by investment, airlift, accommodation, and labour. Limited opportunities for locally driven business growth creates an off balance for local investors on islands. As a result of this off balance and the advent of globalization, island governments have swayed to support foreign direct investment. Kokkranikal and Baum (2011) have noted that tourism jobs are in the public sector and small-scale tourism businesses such as huts are operated by locals. A lack of local participation in tourism investment has been noted (Weaver, 1998) and an over-reliance on foreign investment in small island states occurs (Boukas & Ziakas, 2013). Tourism infrastructure built at an international standard requires large sums of money, not only in the tourism product itself, but also in the utilities and related infrastructure such as a road network. Lewis-Cameron and Roberts (2010) have suggested the need for more equitable involvement of locals in island tourism development.

The extent to which a tourism market develops on an island is directly related to the involvement of market actors in decision-making. Díaz-Pérez, Bethencourt-Cejas, and Álvarez-González (2005) have suggested an island tourism policy of product diversification to capture the high-expenditure market segments. Such a policy will focus resources on high-earning tourism products and services rather than place unsustainable pressures on island communities. Bringing together a market governance structure in the tourism industry is very difficult because of the primarily different interests between the public and private sectors. Market actors may dominate tourism policy-making (McLeod, Chambers, & Airey, 2018). This has implications as the steering of the tourism sector. Christian (2017) has proposed a private governance arrangement whereby a market actor will self-govern actions to support decent working conditions for workers. Buhalis (1999) has noted the dominance of small businesses in the Greek Islands, and the weaknesses of these enterprises in marketing, managing, and planning. Primarily the private sector's interest is in profitability. Bramwell (2006) has observed that in the case of Malta, the discontinuation of the construction of a hotel was more a concern about profitability than it was about conservation. While profit-making is important for sustainability, it is not always a motive for government action, and therefore governments are often caught between balancing the interests of a local community and market actors.

Market structure governance must be integrative of all aspects of the tourism supply chain from hotelier to taxi driver. Market actors' activities in island tourism development face certain challenges. Andrade and Smith (2020) have pointed out that accommodation providers on the island were obliged to work with intermediaries based on access to airlift. The power dynamics among market actors contribute to the outcomes of the tourism industry that eventually develops. Hampton and Christensen (2007) have considered the competing industries in islands and the resultant resource competition, and crowding-out of capital for tourism activities. While destination marketing is an activity for national tourism organizations, in some instances, such activities may be taken over by market actors. Governments may leave the governance of the tourism sector to market actors through an arrangement of a public–private partnership. Chaperon (2017) have noted the role of market actors as a governance mechanism in island tourism, and the benefits were greater efficiencies, and expertise with a public–private partnership arrangement.

6.4 Island tourism associations, networks, and communities

Associations, networks, and communities are the informal governance processes in tourism (Hall, 2011). Tourism associations operate at the local level and conjoin at the regional level with associations in other islands. Two main regional tourism associations with islands are the Caribbean Tourism Organization (Caribbean Tourism Organization, 2023) and the Pacific Asia Travel Association (Pacific Asia Travel Association, 2023). Island tourism networks are of varying sizes depending on the number of tourism actors in the industry. Farmaki (2015) has noted the network-based approach to tourism governance structures supported by the decentralization of tourism governance. Based on the informality of associations, networks, and community organizations, the sustainable governance of these networks must be understood. Nevertheless, collaboration within these informal network structures occurs for several reasons. Keyim (2018) has suggested that collaborative governance results in socio-economic benefits to the community. In small islands, community involvement will allow local residents to participate in the tourism value chain and create greater economic linkages between the tourism industry and the local community (Walker, Lee, & Li, 2021). A meta-governance arrangement is one way to deal with the inadequacies of governance in island destinations. Amore and Hall (2016) have proposed meta-governance to deal with the failure of governance arrangements as the role of the state in tourism development has been challenged by changing power dynamics.

Island tourism networks require a consensus that aligns with the vision and mission of the tourism sector. Tourism industry stakeholders come together to achieve a certain purpose for the growth and development of tourism. Guided by a supporting policy framework, decisions must be taken for the benefit of sustainable island tourism development. Collaboration is essential for island tourism growth. Contributing to sustainability practices on an island such as Gili Trawangan required a multi-stakeholder partnership and collaboration among stakeholders (Graci, 2016). Tourism networks as governance structures must be built up (Dredge, 2003). Steering a directional path of the tourism sector depends on the composition of actors involved in the policy decisions of a tourism sector. With dominant market actors, the tourism sector will take on a development path that benefits those actors, and with dominant community-focused actors then the tourism sector development will serve the interests of local communities (McLeod et al., 2018). Awareness, community unity, and power are key pillars for public participation in tourism efforts and the starting point for community integration in tourism development (Mitchell & Reid, 2001). Communities and local participation are important for sustainable tourism development in islands as local investors with historic family and sociocultural ties are integrated into island communities. Foreign investors may not have such ties. Bichler (2021) has argued that local resident participation is imperative for effective tourism governance. A case study about the role of non-governmental and non-market actors in the sustainable development of the Galápagos Islands is included to explore the role of non-profit organizations.

Mini Case Study 5: Non-profit organizations and sustainable development of the Galápagos Islands

The Galápagos Islands was designated a World Heritage Site and protected for its natural assets (International Galapagos Tour Operators Association, 2023). The Galapagos Islands were the site for the observations of Charles Darwin's theory of evolution (Burbano & Meredith, 2021). A national park was created by executive decree in 1959, and the Galápagos National Park comprises 330 islands, islets, and rocks, and zones include, absolute protection zone, ecosystem conservation and restoration zone, and impact reduction zone (Government of the Republic of Ecuador, 2023). A non-profit organization of travel companies, conservation organizations, and other groups has called for a limit on tourists visiting the islands. The International Galápagos Tourism Operators Association called for a limit to land-based tourism based on a 59% increase from 170,000 in 2010 to 270,000 in 2019 (Edenedo, 2023). The Galápagos Conservancy, a US-based non-profit, is set up to protect and restore the more than 2,000 unique species on the Galápagos Islands (Galápagos Conservancy, 2023).

Mini Case Study 5 Discussion Questions

1 Construct a diagram showing the contributions of the non-profit organizations in the Galápagos Islands.
2 Explain the challenges of tourism governance on the Galápagos Islands.
3 Non-profit organizations are critical for sustainable tourism development. Discuss.

6.5 Conclusion

This chapter discusses the destination organization and stakeholders involved with island tourism sustainable development. The involvement of government and market actors is clearer than the role of non-government or non-profit actors in the governance of island tourism sustainable development. Islands can strengthen tourism governance by considering different actor-led approaches to tourism governance that will result in tourism benefits to residents and communities. Tourism policies that address governance structures must consider the equality of tourism benefits (Boukas & Ziakas, 2016). Tourism stakeholders contribute to decision-making and policy-making. Governance of island tourism must be based on the interactions of stakeholders and therefore a collaborative governance approach is preferred.

Chapter 6 discussion questions

1 Compare and contrast the tourism stakeholders in three island destinations.
2 Discuss the evolution of island destination organizations. In your discussion, identify the factors that contribute to that evolution.
3 Evaluate the roles of island tourism stakeholders in sustainable development.

References

Alipour, H., & Kilic, H. (2005). An institutional appraisal of tourism development and planning: The case of the Turkish Republic of North Cyprus (TRNC). *Tourism Management, 26*(1), 79–94.

Amoamo, M. (2013). Development on the periphery: A case study of the sub-national island jurisdiction of Pitcairn Island. *Asia Pacific Viewpoint, 54*(1), 91–108.

Amore, A., & Hall, C. M. (2016). From governance to meta-governance in tourism? Re-incorporating politics, interests and values in the analysis of tourism governance. *Tourism Recreation Research, 41*(2), 109–122.

Andrade, A., & Smith, K. A. (2020). Tourism distribution in small island destinations: The case of Fernando de Noronha, Brazil. *Journal of Hospitality and Tourism Insights, 3*(2), 171–189.

Bahamas Hotel & Tourism Association (2024). Bahamas Hotel and Tourism Association: The unified voice of Bahamian Tourism. Retrieved from www.bhahotels.com/

Baldacchino, G. (2004). Autonomous but not sovereign?: A review of island sub-nationalism. *Canadian Review of Studies in Nationalism, 31*(1/2), 77–91.

Becken, S., & Loehr, J. (2022). Tourism governance and enabling drivers for intensifying climate action. *Journal of Sustainable Tourism*, 1–19. doi:10.1080/09669582.2022.2032099

Bichler, B. F. (2021). Designing tourism governance: The role of local residents. *Journal of Destination Marketing & Management, 19*(March), 100389.

Boukas, N., & Ziakas, V. (2013). Impacts of the global economic crisis on Cyprus tourism and policy responses. *International Journal of Tourism Research, 15*(4), 329–345.

Boukas, N., & Ziakas, V. (2016). Tourism policy and residents' well-being in Cyprus: Opportunities and challenges for developing an inside-out destination management approach. *Journal of Destination Marketing & Management, 5*(1), 44–54.

Bramwell, B. (2006). Actors, power, and discourses of growth limits. *Annals of Tourism Research, 33*(4), 957–978.

Britton, S. G. (1982). The political economy of tourism in the Third World. *Annals of Tourism Research, 9*(3), 331–358.

Buhalis, D. (1999). Tourism on the Greek Islands: Issues of peripherality, competitiveness and development. *International Journal of Tourism Research, 1*(5), 341–358.

Burbano, D. V., & Meredith, T. C. (2021). Effects of tourism growth in a UNESCO World Heritage Site: Resource-based livelihood diversification in the Galapagos Islands, Ecuador. *Journal of Sustainable Tourism, 29*(8), 1270–1289.

Caribbean Tourism Organization (2023, January 13, 2024). About CTO. Retrieved from www.onecaribbean.org/about-cto/

Chaperon, S. (2017). Tourism industry responses to public–private partnership arrangements for destination management organisations in small island economies: A case study of Jersey, Channel Islands. *International Journal of Tourism Policy, 7*(1), 23–41.

Christian, M. (2017). Protecting tourism labor? Sustainable labels and private governance. *GeoJournal, 82*(4), 805–821.

Díaz-Pérez, F., Bethencourt-Cejas, M., & Álvarez-González, J. (2005). The segmentation of canary island tourism markets by expenditure: Implications for tourism policy. *Tourism Management, 26*(6), 961–964.

Douglas, C. H. (2006). Small island states and territories: Sustainable development issues and strategies – Challenges for changing islands in a changing world. *Sustainable Development, 14*(2), 75–80.

Dredge, D. (2003). *Networks, local governance and tourism policy: Local tourism associations under the microscope.* Paper presented at the CAUTHE 2003, Lismore, NSW.

Edenedo, N. (2023, August 4, 2023). Tour operators call for limits on Galapagos land visitors. *Travel Weekly.* Retrieved from www.travelweekly.com/Travel-News/Tour-Operators/Tour-operators-call-for-limits-on-Galapagos-visitors

Farmaki, A. (2015). Regional network governance and sustainable tourism. *Tourism Geographies, 17*(3), 385–407.

Farsari, I. (2021). Exploring the nexus between sustainable tourism governance, resilience and complexity research. *Tourism Recreation Research, 48*(3), 352–367.

Fiji Hotel and Tourism Association (2024). About FHTA. Retrieved from https://fhta.com.fj/about-fhta/

Galápagos Conservancy (2023). Our mission. Retrieved from www.galapagos.org/about_us/about-us/

Galápagos Conservation Trust (2024). The Galápagos Islands need your help. Retrieved from https://galapagosconservation.org.uk/

Government of the Republic of Ecuador (2023). Galápagos National Park. Retrieved from https://galapagos.gob.ec/parque-nacional-galapagos/

Government of the Republic of Ecuador (2024). Governing council of the special regime of Galápagos. Retrieved from www.gobiernogalapagos.gob.ec/

Graci, S. (2008). What hinders the path to sustainability. A study of barriers to sustainable tourism development in Gili Trawangan, Indonesia. *Pacific News, 29*(January/February), 28–31.

Graci, S. (2016). Collaboration and partnership development for sustainable tourism. In K. Meyer-Arendt & A. A. Lew (Eds.), *Understanding Tropical coastal and island tourism development* (pp. 25–42). Abingdon: Routledge.

Hall, C. M. (2011). A typology of governance and its implications for tourism policy analysis. *Journal of Sustainable Tourism, 19*(4–5), 437–457.

Hampton, M. P., & Christensen, J. (2007). Competing industries in islands a new tourism approach. *Annals of Tourism Research, 34*(4), 998–1020.

Higgins-Desbiolles, F. (2022). The ongoingness of imperialism: The problem of tourism dependency and the promise of radical equality. *Annals of Tourism Research, 94*(May), 103382.

International Galapagos Tour Operators Association (2023). Galápagos at risk. Retrieved from www.igtoa.org/galapagos_at_risk

International Galapagos Tour Operators Association (2024). About IGTOA. Retrieved from www.igtoa.org/about

Ivanka, H. G., Marion, K., Anthony, W. I., & Rob, L. (2023). Tourism destination research from 2000 to 2020: A systematic narrative review in conjunction with bibliographic mapping analysis. *Tourism Management, 95*(April), 104686.

Kerr, S. A. (2005). What is small island sustainable development about? *Ocean & Coastal Management, 48*(7–8), 503–524.

Keyim, P. (2018). Tourism collaborative governance and rural community development in Finland: The Case of Vuonislahti. *Journal of Travel Research, 57*(4), 483–494.

Klint, L. M., Wong, E., Jiang, M., Delacy, T., Harrison, D., & Dominey-Howes, D. (2012). Climate change adaptation in the Pacific Island tourism sector: Analysing the policy environment in Vanuatu. *Current Issues in Tourism, 15*(3), 247–274.

Kokkranikal, J., & Baum, T. (2011). Tourism and sustainability in the Lakshadweep Islands. In J. Carlsen & R. Butler (Eds.), *Island tourism: sustainable perspectives* (pp. 54–71). Wallingford, UK: CABI.

Lewis-Cameron, A., & Roberts, S. (2010). Small island developing states: Issues and prospects. In S. Roberts & A. Lewis-Cameron (Eds.), *Marketing island destinations: Concepts and cases* (pp. 1–10). London: Elsevier.

Liasidou, S. (2019). Understanding tourism policy development: A documentary analysis. *Journal of Policy Research in Tourism, Leisure and Events, 11*(1), 70–93.

Mallorca Sustainable Tourism Observatory (2024a). The agency for tourism strategy of the Balearic Islands. Retrieved from https://stomallorca.com/en/entidades-publicas/

Mallorca Sustainable Tourism Observatory (2024b). Hotel Business Federation of Mallorca. Retrieved from https://stomallorca.com/en/entidades-privadas/

Marilles Foundation (2024). Marilles Foundation. Retrieved from https://marilles.org/en

McLeod, M. (2022). Tourism destination recovery, a case study of Grand Bahama Island. In I. Bethell-Bennett, S. Rolle, J. Minnis, & F. Okumus (Eds.), *Pandemics, disasters, sustainability, tourism* (pp. 93–108). Leeds, UK: Emerald Publishing.

McLeod, M., & Airey, D. (2007). The politics of tourism development: A case of dual governance in Tobago. *International Journal of Tourism Policy, 1*(3), 217–231.

McLeod, M., Chambers, D., & Airey, D. (2018). A comparative analysis of tourism policy networks. In M. McLeod & R. Croes (Eds.), *Tourism management in warm-water island destinations* (pp. 77–94). Wallingford, UK: CABI.

Ministry of Trade Co-operatives Small and Medium Enterprises (2024). Division of tourism. Retrieved from https://mcttt.gov.fj/division/tourism/

Mitchell, R. E., & Reid, D. G. (2001). Community integration: Island tourism in Peru. *Annals of Tourism Research, 28*(1), 113–139.

NatureFiji (2024). NatureFiji about us. Retrieved from https://naturefiji.org/about-us/

Organization for Responsible Governance (2023). The Organization for Responsible Governance: Uniting Bahamians toward a brighter future. Retrieved from www.orgbahamas.com/

Pacific Asia Travel Association (2023, January 13, 2024). About PATA. Retrieved from www.pata.org/about-pata

Partelow, S., & Nelson, K. (2020). Social networks, collective action and the evolution of governance for sustainable tourism on the Gili Islands, Indonesia. *Marine Policy, 112*(February). doi:10.1016/j.marpol.2018.08.004

Peterson, R. R. (2020). Over the Caribbean top: Community well-being and over-tourism in small island tourism economies. *International Journal of Community Well-Being, 6*(November), 1–38.

Said, F., & Farid, R. S. (2022). Marine Tourism sustainability strategy with triple helix support: a case study of West Sulawesi marine tourism. *Journal of Humanities and Social Sciences Studies, 4*(4), 330–335.

Scheyvens, R., & Momsen, J. (2008). Tourism in small island states: From vulnerability to strengths. *Journal of Sustainable Tourism, 16*(5), 491–510.

Seychelles Hospitality and Tourism Association (2024). Seychelles Hospitality and Tourism Association. Retrieved from www.facebook.com/p/Seychelles-Hospitality-and-Tourism-Association-100064933020394/

Seychelles Islands Foundation (2024). About. Retrieved from www.sif.sc/about

The Bahamas Department of Aviation (2023). *National aviation strategic three-year plan.* The Bahamas: The Bahamas Ministry of Tourism, Investments & Aviation. Retrieved from https://tempo.cdn.tambourine.com/bmot-aviation/media/nasp-powerpoint-presentation_july-2023_ver2-3-compressed-64c13a4a51999.pdf

The Bahamas Ministry of Tourism (2024). Bahamas Ministry of Tourism, Investments & Aviation launches first aviation week. Retrieved from www.bahamas.com/pressroom/bahamas-ministry-of-tourism-investments--aviation-launches-first-aviation-week

Tourism Seychelles (2024). Official tourism department website: Tourism Seychelles. Retrieved from https://tourism.gov.sc/

Uysal, M., & Modica, P. (2016). Island tourism: Challenges and future research directions. In P. Modica & M. Uysal (Eds.), *Sustainable island tourism: Competitiveness and quality of life* (pp. 173–188). Wallingford, UK: CABI.

Valdivielso, J., & Moranta, J. (2019). The social construction of the tourism degrowth discourse in the Balearic Islands. *Journal of Sustainable Tourism, 27*(12), 1876–1892.

Walker, T. B., Lee, T. J., & Li, X. (2021). Sustainable development for small island tourism: Developing slow tourism in the Caribbean. *Journal of Travel & Tourism Marketing, 38*(1), 1–15.

Wan, Y. K. P., Li, X., Lau, V. M.-C., & Dioko, L. D. (2022). Destination governance in times of crisis and the role of public–private partnerships in tourism recovery from COVID-19: The case of Macao. *Journal of Hospitality and Tourism Management, 51*(June), 218–228.

Weaver, D. B. (1998). Peripheries of the periphery: Tourism in Tobago and Barbuda. *Annals of Tourism Research, 25*(2), 292–313.

7   Tourism governance systems

7.1 Introduction

A tourism governance system suggests that the components of the tourism sector work together in a manner that is beneficial to the tourism stakeholders. Governance systems are set up to implement and monitor island tourism activities (McLeod, Dodds, & Butler, 2021). Governance is the setting up of policies, regulations, enforcements to keep tourism interests on track to achieve a stated purpose. Justifiably, governance involves a multitude of tourism actors, and understanding a governance outcome requires systems thinking. Rodriguez-Giron and Vanneste (2019) have detailed an integrated tourism systems thinking conceptual framework and have stated the system components, system structure, and systems properties. As it relates to the nature of governance of island tourism, some conceptualization of the components is needed as the inputs and the processes of a system determine the outputs, and its outcomes. In tourism governance, application of the Rodriguez-Giron and Vanneste (2019) conceptual framework suggests that outcomes are emergent, self-organized, or non-linear based on the system components (elements, function, dynamics, and environment) and structure (abstraction and hierarchy). Realistically, the multiplicity of elements, actors, both internal and external, involved with tourism development and management on an island are difficult to identify and map. Baggio (2014) has suggested that governance actions need to be aligned based on unpacking the complexity of the tourism destination system. Some degree of measurement using sophisticated methodologies is required. Nonetheless, a step-by-step approach that unpacks the governance components that contribute to sustainable island tourism development and how these contribute to good governance of island destinations may be done.

Governance is sometimes referred to as 'good' governance. In small island developing states, a lack of state support for tourism development may result in poor governance (Sharpley & Ussi, 2014). Herein is the importance of 'good' governance, meaning that the governance system works for the better good. The question may be governance for who. If an island tourism governance system works, then all will benefit including locals, tourists, governments, businesses, and communities. Varying perspectives have emerged about whom the tourism activities benefit. To

DOI: 10.4324/9781003435112-9

the extent that there are suggestions that tourism does not benefit locals, then some aspect of governance may not be working. Good governance suggests that the right set of benefits are obtained from tourism activities on an island. Benefits must outweigh costs for island tourism development to be sustainable. Xing and Dangerfield (2018) have considered mass tourism in island destinations and have noted that sustainable development issues are related to system problems, and understanding and making coordinated changes in the system are needed for effective effects of sustainable tourism development policies. This chapter unpacks some of the major governance challenges in island tourism. Such an endeavour facilitates the identification of potential frameworks and models that can handle these challenges. Lastly, a sustainable tourism governance system is proposed to maintain beneficial island tourism in the long run.

7.2 Island tourism governance challenges

A governance challenge in tourism may be viewed as an aspect that is difficult to manage or control. Islands are diverse ecosystems with varying political, economic, socio-cultural, and environmental realities. Kurniawan, Adrianto, Bengen, and Prasetyo (2019) have proposed the application of a socio-ecological system approach for understanding sustainable island tourism development. Social and ecological systems suggest that human and natural aspects of islands are intertwined, and impacts occur as the system comes out of balance. A multiplicity of aspects within these dimensions affects island tourism and for the purpose of exploring governance challenges, only the major challenges have been outlined. Based on the literature, major challenges affecting island tourism ecosystems include forging an independent path for development, climate change and crisis, beaches and marine ecosystem sustainability, biodiversity loss and pollution, funding and financial aspects, migration and limited opportunities, land ownership and tenure, and overall infrastructure for development including water and electricity (Harrison & Pratt, 2015; McLeod, 2022; Peterson, 2020; Walker & Lee, 2022; Wolf et al., 2022).

Many islands were colonized, and island governments still struggle with self-determination and creating an independent path for development. Colonial legacies, administrations, and decisions continue to affect island tourism governance (Harrison, 2001; Rao, 2002). Governance practices are modelled after political directorates, and such directorates are placed within an international community that impose conditionalities. As such island government's freedom to choose certain courses of action is constrained. Islands and the prevailing tourism industries are affected by colonial histories and legacies, inequalities and imperialistic notions by foreign dominance of tourism businesses (Higgins-Desbiolles, 2022). Governance systems are built on existing conditions that may not realize beneficial tourism.

Climate change is the greatest crisis of the 21st century, and islands are the most affected destinations. Critically, Klint et al. (2012) have viewed the small size of islands as an advantage in dealing with the climate change crisis. Contrarily, governments and residents on small islands have made plans to leave and hence the term *climate refugee* (Farbotko, 2010). The term climate refugee has taken on a new meaning, as residents on islands face the stark reality of fleeing their homes, environments, and communities as the very existence of their island is threatened by rising sea levels and disasters. The movement of people because of the impacts of climate change requires in-depth research on the social and cultural adjustments in the emigrated countries. Wolf et al. (2022) have noted that more work is needed to explore climate adaptation in small island developing states. A challenge is the playing down of perceived risks as countries such as the Maldives seek to attract investment and tourists (Shakeela & Becken, 2015).

Development of infrastructure is a governance issue on islands. Island infrastructure in most cases may not withstand the strongest hurricane or cyclone (McLeod, 2022). Infrastructure is old, outdated, and often inadequate to handle the pressure of population growth and disasters or emergencies. Rodríguez, Parra-López, and Yanes-Estévez (2008) have noted the tourism pressures on existing infrastructure in Tenerife and have suggested that new infrastructure provides economic benefits. While development of new infrastructure on old infrastructure may result in dual maintenance systems, Scheyvens and Momsen (2008) have noted a space constraint on an island as one of the reasons for limited infrastructure and economic opportunities. Governance of infrastructure across several islands is another issue. Outer islands' infrastructure is often poor when compared with the main islands in an archipelago (Connell, 2016). The provision of good health and education systems, good transport, air- and seaports' infrastructure are critical for islands and these services support a population that may not have access to such resources otherwise.

Marine ecosystem sustainability, including issues about biodiversity, beaches, and pollution, is critical to support island tourism activities (Kokkranikal & Baum, 2011; Kurniawan et al., 2019). Temperature changes are affecting marine ecosystems. The threat of the sargassum seaweed in Caribbean tourism destinations has resulted in a loss of livelihood in several communities (Origin by Ocean, 2023). Natural disasters related to climate impacts have received calls for better facilities on islands to handle these threats. The use of marine ecosystems from the point of view of a free tourism resource has been debated, and whether marine ecosystems provide leisure activities or are used for food and welfare purposes, tourism activities require these resources for beneficial purposes. The loss of coral reefs and mangroves, changes in species, and the environmental impact of cruises, shipping, and other vessels, must be monitored and controlled (Mycoo, 2006). As a governance strategy, island governments have begun to focus on the Blue Economy and have adopted strategies to grow economic potential from marine resources on, in, and below the seabed (Sammler, 2016).

With limited economic opportunities for growth and development, island governments have funding and financial challenges. One approach to increase financial strength has been to diversify an island economy. Diversification into offshore financial houses competes with the tourism sector for resources

(Hampton & Christensen, 2007). Financing tourism activities requires funding from several sources, as tourism activities have financial impacts (Moghal & O'Connell, 2018). Governance of island revenues poses another challenge when tourism earnings are not reinvested in the development of the tourism sector. Funding climate change adaptation is another complexity that poses another financing challenge. Climate financing in small island developing sates is needed, and the Bridgetown declaration is an important initiative to handle island sustainability (Wolf et al., 2022).

Island tourism development has land tenure and ownership challenges. Land ownership can be communal ownership as in Melanesia territories or small holdings as in French Polynesia (Harrison & Pratt, 2015). Non-transfer of family land ownership to descendants has consequences for generational wealth. Local rights to land ownership have to be addressed using a legislative framework for greater participation of local residents in the tourism industry (Scheyvens & Momsen, 2008). In addition, land tenure issues, laws, and regulations affect land resources for tourism development. Weaver (1998) has noted the changing of land ownership to second homeowners as a means of foreign investment. A consequence of such a policy has been the exorbitant property prices that are unaffordable for locals (Bardolet & Sheldon, 2008).

7.3 Island tourism governance frameworks

An island tourism governance framework supports the myriad of challenges island governments and communities must handle. Several frameworks and models were considered and the most relevant ones for the sustainable development of island tourism have been outlined. Klint et al. (2012) have recommended adaptive strategies to reduce climate change impacts. An adaptation framework for climate change is constrained by institutional and governance arrangements, and adjustments are needed to manage one of the greatest challenges facing small island developing states (Robinson, 2018, 2020). Adaptative governance has to be an overarching framework that guides building a resiliency framework for island tourism destinations (McLeod, 2020). Sustainable governance can be proposed as a framework. Roxas, Rivera, and Gutierrez (2020) have proposed an ecotourism sustainability framework using systems thinking and have argued that ecotourism contributes to profitable businesses, creation of local jobs, and conservation.

Building a green economy supports green tourism practices. A green economy framework allows islands to manage resources in different ways. Investments in a green economy have at the helm the creation of environmentally sustainable practices. Focusing on low carbon emissions, green economy practices may be adopted to realize sustainability goals (Pan et al., 2018). Principles of internal resource utilization assist with building island sustainability. In reference to Guam, circular economy principles that reduce wastage of resources can be applied in managing the environmental challenges of small island states (Schumann, 2020). Utilizing local resources also benefits tourism. Yfantidou and Matarazzo (2017) have referred to 'Green tourism' and have applied this concept to developing countries to enhance the contribution tourism makes to local economies and the maintenance of economic benefits. Green economy growth is balanced growth that supports inclusion, consultation, accountability, and openness (Fay, 2012).

Islands are endowed with blue economy resources. The blue economy relates to the economics of oceans and seas that are managed for the economic welfare of residents. Remote islands in the South Pacific are surrounded by vast blue economy resources that are under-utilized. Dwyer (2018) has pointed out that small island developing states need to understand the complexities of governance issues of oceans and integrated governance of coastal and marine tourism. Picken (2023) has suggested that blue economy governance involves multilateral governance. Governance principles of the blue economy include balanced economic growth, social equity, and environmental preservation (Phelan, Ruhanen, & Mair, 2020). Blue economy as a governance approach can add value to island tourism, but these resources must be carefully managed. Alipour and Arefipour (2020) have proposed co-management of Common Pool Resources for Northern Cyprus to allow for a multi-level governance structure with public, private, and non-governmental organizations working together to share power and collaborate. The zoning of marine resources for use in the blue economy enriches island governments and communities and supports economic diversification and funds tourism development. One example of this is the creation of exclusive economic zones for the extraction of ocean mineral resources in the South Pacific (Sammler, 2016). To reduce overfishing, governance of marine resources also involves finding alternative sources of income, such as beekeeping, for residents of fishing communities. A blue economy governance framework contributes to addressing conservation challenges on islands.

Integrated governance can support island tourism sustainability. First, the elements of an island tourism system must be understood. Mai and Smith (2018) have used scenario-based planning to conduct system dynamic modelling to understand island tourism sustainability. Second, the resources to be managed must be set out. The Caribbean Tourism Organization proposed a tourism policy and sustainable development framework that addresses capacity management, destination management, marketing and public relations, and risk management (Caribbean Tourism Organization, 2020). Third, an approach for the sustainable management of community resources must be developed. While the process and implementation of community-based tourism initiatives have had some success (Kontogeorgopoulos, Churyen, & Duangsaeng, 2014), the size of investment required and the capacity of the communities involved are constraints with adapting this approach to solve some of the tourism governance issues (Goodwin, 2002). Fourth, integrated governance assists with implementation. In Samoa, integrated governance between the government system and village system encourages the implementation of environmental regulations (Jiang & DeLacy, 2014). A governance framework is needed to manage the sargassum seaweed problem in the Caribbean. A mini case study follows to guide discussion about governance systems.

Mini Case Study 6: The sargassum seaweed problem in the Dominican Republic

The Dominican Republic is located on the Hispaniola Island with Haiti and is part of the Greater Antilles and is one of the top five Caribbean tourism destinations. According to the minister of tourism, the period from January to July 2023 recorded visitor arrivals of more than 6.2 million (*Dominican Today*, 2023). Sargassum seaweed originate from the Sargasso Sea in the Atlantic Ocean and washes onshore a beach (Figure 7.1). Beaches are a major attraction of Caribbean destinations. The presence of the sargassum seaweed on the beach is an environmental hazard based on the gases that are emitted. The sargassum seaweed has plagued the coastline of some Caribbean destinations since 2011 (Lennon, 2022). Some of challenges for addressing the sargassum issue include understanding the causes, the health issues associated with the seaweed, and funding the clean-up (Lennon, 2022). The management of the sargassum seaweed problem is an ongoing issue in the Dominican Republic. Several partners have come together to provide solutions and solve this problem. One example of such a partnership is among Origin by Ocean, SOS Carbon, NODO Logistics, and Grupo Puntacana (Origin by Ocean, 2023).

Figure 7.1 Sargassum seaweed on a beach.

Source: Origin by Ocean (2023).

Mini Case Study 6 discussion questions

1 Explain the issues about the sargassum seaweed in the Dominican Republic.
2 Apply a governance framework for the management of the sargassum seaweed.
3 Discuss the pros and cons of a multi-faceted partnership to solve problems such as the sargassum seaweed.

7.4 Island tourism sustainable governance

Sustainable governance indicators can contribute to understanding island tourism sustainable development. Policy performance, democracy, and executive capacity and accountability are the key sustainable governance indicators (Stiftung, 2011). Given the governance challenges and the governance frameworks, application of sustainable governance means that the governance activities are working to resolve governance challenges. Based on the size of an island's economy, sustainable governance seems feasible as the number of actors and activities are substantially less than larger economies. Kelman (2019) has critiqued sustainability initiatives for island tourism and has suggested that no real contribution may be made towards sustainability but rather sustainability labels are being applied. Such a suggestion means that efforts toward sustainable development needs to be addressed in island economies. Alipour, Vaziri, and Ligay (2011) have recommended institutional restructuring to obtain closer cooperation between the public and private sector. Public–private partnerships are an effective means to support sustainable governance as such an arrangement builds governance capacity. Strong partnerships between the public and private sectors facilitate good governance practices as trust built between these sectors enhance working relationships and achieve results.

The creation of effective sustainable development policies and monitoring of policy performance are essential for sustainable development. The changing world system and external shocks mean that policies are needed to support sustainable development. Boukas and Ziakas (2016) have pointed out in the case of Cyprus that small island states have a level of foreign dependency that make these countries vulnerable to global shocks and crises. Sustainable use of resources and inclusive growth as a policy in island tourism are key island tourism governance systems (Le, 2023). For the creation of policies, policy direction on the pooling of limited resources and adoption of governance processes are needed. For the monitoring of policies, the use of data for policy performance can greatly enhance a sustainable governance system. Greater use of information and communication technologies contributes to smart specialization in small island states and can enhance sustainability indicators (Bhaduri & Pandey, 2020).

Democracy is a pillar of sustainable governance (Croissant & Pelke, 2022). The adoption of good governance principles benefit sustainable island tourism governance systems (Sharpley & Ussi, 2014). Participatory, collaborative, and consensus building approaches to governance are effective in sustaining tourism activities. Sustainable governance occurs at all levels within a governance eco-system, national, regional, and local, and helps with the effectiveness of good gov-ernance. Saarinen and Gill (2018) have argued that better governance in tourism builds adaptive capacity and strengthens resilience. The Cittàslow movement and slow tourism that focuses on local governance contributes to island tourism sus-tainability (Walker & Lee, 2022). The setting up of a regulatory framework that has overall buy-in and agreement by those affected will facilitate greater sustain-ability of governance actions. Transparency, accountability, and responsiveness to changes in government actions aid in acceptance of those changes and sustain-ability in governance practices. With regards to climate change adaptation, Wong, Jiang, Klint, Dominey-Howes, and DeLacy (2013) have pointed out the lack of transparency about the implementation of policies. A governance process must be transparent to be effective in realizing tourism goals, and efficient given time and space constraints in island tourism.

7.5 Conclusion

This chapter sets out the key aspects that contribute to island tourism governance systems, including the challenges, frameworks, and sustainable governance. The multiplicity of challenges faced by island governments building and managing a successful tourism sector requires frameworks that support sustainable develop-ment. The chapter proposed the blue economy and green economy, community-based, and integrated development frameworks. Integrated development as a governance framework and sustainable governance as a governance process can go a long way in resolving governance challenges in island tourism. Sustainable development must also be supported by sustainable governance within blue economy and green economy frameworks. The chapter did not address institu-tional aspects of governance as an approach, rather the selection of appropriate frameworks to address goverance for sustainability and the adoption of good gov-ernance principles were explored.

Chapter 7 discussion questions

1 Explore the governance challenges discussed in the chapter and construct a diagram that sets out these challenges in any island destination.
2 Apply a blue economy and green economy framework to the governance practices of an island destination.
3 Discuss the barriers of sustainable governance practices in island destinations.

References

Alipour, H., & Arefipour, T. (2020). Rethinking potentials of co-management for sustainable common pool resources (CPR) and tourism: The case of a Mediterranean island. *Ocean & Coastal Management, 183*(January), 104993.

Alipour, H., Vaziri, R. K., & Ligay, E. (2011). Governance as catalyst to sustainable tourism development: Evidence from North Cyprus. *Journal of Sustainable Development, 4*(5), 32–49.

Baggio, R. (2014). Complex tourism systems: A visibility graph approach. *Kybernetes, 43*(3/4), 445–461.

Bardolet, E., & Sheldon, P. J. (2008). Tourism in archipelagos: Hawai'i and the Balearics. *Annals of Tourism Research, 35*(4), 900–923.

Bhaduri, K., & Pandey, S. (2020). Sustainable smart specialisation of small-island tourism countries. *Journal of Tourism Futures, 6*(2), 121–133.

Boukas, N., & Ziakas, V. (2016). Tourism policy and residents' well-being in Cyprus: Opportunities and challenges for developing an inside-out destination management approach. *Journal of Destination Marketing & Management, 5*(1), 44–54.

Caribbean Tourism Organization (2020). Caribbean sustainable tourism policy and development framework. Retrieved from Barbados: https://ourtourism.onecaribbean.org/resour ces/caribbean-sustainable-tourism-policy-framework-2020/

Connell, J. (2016). Competing islands? The Mamanuca and Yasawa Islands, Fiji. In G. Baldacchino (Ed.), *Archipelago tourism* (pp. 182–197). Abingdon: Routledge.

Croissant, A., & Pelke, L. (2022). Measuring policy performance, democracy, and governance capacities: A conceptual and methodological assessment of the sustainable governance indicators (SGI). *European Policy Analysis, 8*(2), 136–159.

Dominican Today (2023). Dominican Republic smashes record in arrivals with 6.2 million visitors through July. *Dominican Today.*

Dwyer, L. (2018). Emerging ocean industries: Implications for sustainable tourism development. *Tourism in Marine Environments, 13*(1), 25–40.

Farbotko, C. (2010). Wishful sinking: Disappearing islands, climate refugees and cosmopolitan experimentation. *Asia Pacific Viewpoint, 51*(1), 47–60.

Fay, M. (2012). *Inclusive green growth: The pathway to sustainable development.* Washington, DC: World Bank Publications.

Goodwin, H. (2002). Local community involvement in tourism around national parks: Opportunities and constraints. *Current Issues in Tourism, 5*(3–4), 338–360.

Hampton, M. P., & Christensen, J. (2007). Competing industries in islands a new tourism approach. *Annals of Tourism Research, 34*(4), 998–1020.

Harrison, D. (2001). Islands, image and tourism. *Tourism Recreation Research, 26*(3), 9–14.

Harrison, D., & Pratt, S. (2015). Tourism in Pacific island countries: Current issues and future challenges. In S. Pratt & D. Harrison (Eds.), *Tourism in Pacific Islands: Current issues and future challenges* (pp. 3–21). Abingdon: Routledge.

Higgins-Desbiolles, F. (2022). The ongoingness of imperialism: The problem of tourism dependency and the promise of radical equality. *Annals of Tourism Research, 94*(May), 103382.

Jiang, M., & DeLacy, T. (2014). A climate change adaptation framework for Pacific Island tourism. In T. Delacy, M. Jiang, G. Lipman, & S. Vorster (Eds.), *Green growth and travelism: Concept, policy and practice for sustainable tourism* (pp. 225–240). Abingdon: Routledge.

Kelman, I. (2019). Critiques of island sustainability in tourism. *Tourism Geographies, 23*(3), 397–414.

Klint, L. M., Wong, E., Jiang, M., Delacy, T., Harrison, D., & Dominey-Howes, D. (2012). Climate change adaptation in the Pacific Island tourism sector: Analysing the policy environment in Vanuatu. *Current Issues in Tourism, 15*(3), 247–274.

Kokkranikal, J., & Baum, T. (2011). Tourism and sustainability in the Lakshadweep Islands. In J. Carlsen & R. Butler (Eds.), *Island tourism: sustainable perspectives* (pp. 54–71). Wallingford, UK: CABI.

Kontogeorgopoulos, N., Churyen, A., & Duangsaeng, V. (2014). Success factors in community-based tourism in Thailand: The role of luck, external support, and local leadership. *Tourism Planning & Development, 11*(1), 106–124.

Kurniawan, F., Adrianto, L., Bengen, D. G., & Prasetyo, L. B. (2019). The social-ecological status of small islands: An evaluation of island tourism destination management in Indonesia. *Tourism Management Perspectives, 31*(July), 136–144.

Le, H. N. (2023). Island tourism development for inclusive growth. In A. Morrison & D. Buhalis (Eds.), *Routledge handbook of trends and issues in tourism sustainability, planning and development, management, and technology* (pp. 160–170). Abingdon: Routledge.

Lennon, C. (Producer). (2022, 16 June 2022). *Barbados and the blue economy – Sargassum solutions*. Retrieved from https://news.un.org/en/audio/2022/06/1120482

Mai, T., & Smith, C. (2018). Scenario-based planning for tourism development using system dynamic modelling: A case study of Cat Ba Island, Vietnam. *Tourism Management, 68*(October), 336–354.

McLeod, M. (2020). Tourism governance, panarchy and resilience in the Bahamas. In S. Rolle, J. Minnis, & I. Bethell-Bennett. *Tourism development, governance and sustainability in the Bahamas* (pp. 103–113). Abingdon, UK: Routledge.

McLeod, M. (2022). Tourism destination recovery, a case study of Grand Bahama Island. In I. Bethell-Bennett, S. Rolle, J. Minnis, & F. Okumus (Eds.), *Pandemics, disasters, sustainability, tourism* (pp. 93–108). Leeds, UK: Emerald Publishing.

McLeod, M., Dodds, R., & Butler, R. (2021). Introduction to special issue on island tourism resilience. *Tourism Geographies, 23*(3), 361–370.

Moghal, Z., & O'Connell, E. (2018). Multiple stressors impacting a small island tourism destination-community: A nested vulnerability assessment of Oistins, Barbados. *Tourism Management Perspectives, 26*(April), 78–88.

Mycoo, M. (2006). Sustainable tourism using regulations, market mechanisms and green certification: A case study of Barbados. *Journal of Sustainable Tourism, 14*(5), 489–511.

Origin by Ocean (2023, January 7, 2022). Washing the coasts of the Dominican Republic: Our alliance to rid the Dominican Republic of sargassum seaweed is in action. Retrieved from www.originbyocean.com/blog/washing-the-coasts-of-the-dominican-republic

Pan, S.-Y., Gao, M., Kim, H., Shah, K. J., Pei, S.-L., & Chiang, P.-C. (2018). Advances and challenges in sustainable tourism toward a green economy. *Science of the Total Environment, 635*(September), 452–469.

Peterson, R. R. (2020). Over the Caribbean top: Community well-being and over-tourism in small island tourism economies. *International Journal of Community Well-Being, 6*(November), 1–38.

Phelan, A., Ruhanen, L., & Mair, J. (2020). Ecosystem services approach for community-based ecotourism: Towards an equitable and sustainable blue economy. *Journal of Sustainable Tourism, 28*(10), 1665–1685.

Picken, F. (2023). Tourism and the blue economy. *Tourism Geographies*, 1–9. doi:10.1080/14616688.2023.2291821

Rao, M. (2002). Challenges and issues for tourism in the South Pacific island states: The case of the Fiji Islands. *Tourism Economics, 8*(4), 401–429.

Robinson, S.-A. (2018). Climate change adaptation limits in small island developing states. In W. L. Filho & J. Nalau (Eds.), *Limits to climate change adaptation* (pp. 263–281). Cham: Springer.

Robinson, S.-A. (2020). Climate change adaptation in SIDS: A systematic review of the literature pre and post the IPCC Fifth Assessment Report. *Wiley Interdisciplinary Reviews: Climate Change, 11*(4), 1–21.

Rodríguez, J. R. O., Parra-López, E., & Yanes-Estévez, V. (2008). The sustainability of island destinations: Tourism area life cycle and teleological perspectives. The case of Tenerife. *Tourism Management, 29*(1), 53–65.

Rodriguez-Giron, S., & Vanneste, D. (2019). Tourism systems thinking: Towards an integrated framework to guide the study of the tourism phenomenon. *Tourism Culture & Communication, 19*(1), 1–16.

Roxas, F. M. Y., Rivera, J. P. R., & Gutierrez, E. L. M. (2020). Framework for creating sustainable tourism using systems thinking. *Current Issues in Tourism, 23*(3), 280–296.

Saarinen, J., & Gill, A. M. (2018). Tourism, resilience, and governance strategies in the transition towards sustainability. In J. Saarinen & A. M. Gill (Eds.), *Resilient destinations and tourism* (pp. 15–33). Abingdon: Routledge.

Sammler, K. G. (2016). The deep pacific: Island governance and seabed mineral development. In E. Stratford (Ed.), *Island geographies* (pp. 24–45). Abingdon: Routledge.

Scheyvens, R., & Momsen, J. (2008). Tourism and poverty reduction: Issues for small island states. *Tourism Geographies, 10*(1), 22–41. doi:10.1080/14616680701825115

Schumann, F. R. (2020). Circular economy principles and small island tourism Guam's initiatives to transform from linear tourism to circular tourism. *Journal of Global Tourism Research, 5*(1), 13–20.

Shakeela, A., & Becken, S. (2015). Understanding tourism leaders' perceptions of risks from climate change: An assessment of policy-making processes in the Maldives using the social amplification of risk framework (SARF). *Journal of Sustainable Tourism, 23*(1), 65–84.

Sharpley, R., & Ussi, M. (2014). Tourism and governance in small island developing states (SIDS): The case of Zanzibar. *International Journal of Tourism Research, 16*(1), 87–96.

Stiftung, B. (2011). Sustainable governance indicators 2011: Policy performance and governance capacities in the OECD. Retrieved from Germany: https://api.pageplace.de/prev iew/DT0400.9783867933933_A18798354/preview-9783867933933_A18798354.pdf.

Walker, T. B., & Lee, T. J. (2022). Contributions to sustainable tourism in small islands: An analysis of the Cittàslow movement. In M. McLeod, R. Dodds, & R. Butler (Eds.), *Island tourism sustainability and resiliency* (pp. 54–74). Abingdon: Routledge.

Weaver, D. B. (1998). Peripheries of the periphery: Tourism in Tobago and Barbuda. *Annals of Tourism Research, 25*(2), 292–313.

Wolf, F., Moncada, S., Surroop, D., Shah, K. U., Raghoo, P., Scherle, N., … Havea, P. H. (2022). Small island developing states, tourism and climate change. *Journal of Sustainable Tourism*, 1–19. doi:10.1080/09669582.2022.2112203

Wong, E., Jiang, M., Klint, L. M., Dominey-Howes, D., & DeLacy, T. (2013). Evaluation of policy environment for climate change adaptation in tourism. *Tourism and Hospitality Research, 13*(4), 201–225.

Xing, Y., & Dangerfield, B. (2018). Modelling the sustainability of mass tourism in island tourist economies. In M. Kunc (Ed.), *System dynamics: Soft and hard operational research* (pp. 303–327). London: Springer.

Yfantidou, G., & Matarazzo, M. (2017). The future of sustainable tourism in developing countries. *Sustainable Development, 25*(6), 459–466.

8 Tourism governance strategies

8.1 Introduction

An island tourism governance strategy is a direction to achieve a particular governance action. Governance actions are framed based on destination goals. Island tourism development has been framed using Sustainable Development Goals to build resilience (McLeod, 2023b; Rasoolimanesh, Ramakrishna, Hall, Esfandiar, & Seyfi, 2023). Governance approaches to achieve island tourism resilience and sustainability can be categorized as social, environmental, political, and economic approaches. Social governance approaches contribute to the well-being of society. The environmental, social, and governance (ESG) framework that shifts focus away from economics has gained traction in developing countries. ESG balances the development agenda by selecting projects that will result in a greater benefit overall. Ternel and Greyling (2018), with consideration of the ESG, have suggested that sustainability as a concept in island tourism remains at an embryonic stage. Environmental governance is not a given as it requires a coordinated effort. Said and Farid (2022) explored marine tourism in West Sulawesi, Indonesia, and found that a strong commitment, a comprehensive and coordinated plan, that supports and ensures sustainability is needed. Managing the environment requires a broad range of resources and regulatory controls to enforce environmental laws. Politics and governance in sustainable tourism are related to the exercise of policies that will achieve sustainable tourism development. Several elements are considered to address the issue of governance through the exercise of politics, and finally, economic governance is the way in which economic activities are managed to create a prosperous economy. Governance is an evolving process, and it also incorporates economic, social, political, and environmental governance elements to support sustainable island tourism.

This chapter explores island tourism governance strategies utilizing balanced governance approaches that involve the pillars of sustainability, economic, social, and environmental, while addressing the politics involved in governance practices. Tourism governance poses a transnational challenge as both local and foreign stakeholders are involved (Dredge & Jamal, 2013). Nonetheless, tourism governance has to be framed and provide direction for a tourism industry, and involve all tourism stakeholders including local residents (Bichler, 2021). While

DOI: 10.4324/9781003435112-10

the boundaries between the governance elements may be blurred, as there are interrelationships among the elements, using this approach unpacks considerations and workable practices to institute sustainable development policies for island tourism. A review of practices is particularly important as islands with dominant tourism industries seek to understand the best ways to address the myriad of management and development issues.

8.2 Island tourism governance social strategies

Social governance means that actions benefit an island's society. Social responsibility goes further and suggests that actions are taken based on a moral obligation. In that regard, social responsibility is viewed as a good governance principle. In exploring social governance, the ESG activities may be applied as businesses address relationships with customers, employees, and communities. Social sustainability principles are also applicable to address issues of quality of life, equality, diversity, and cohesion of members in society (Holladay & Powell, 2013). Local communities require greater involvement in tourism activities for the creation of social benefits (Kinseng, Nasdian, Fatchiya, Mahmud, & Stanford, 2018). Parmawati, Pangestuti, Wike, and Hardyansah (2020) have noted the need for increasing human resource capacity, while education and negotiation are key strategies to achieve sustainable tourism in the case of Red Island Beach, Indonesia. For social sustainability of island tourism and equality, social divisions, and class systems that off balance the distribution of wealth must be addressed. Bernardo and Jorge (2019) have considered the negative social effects of tourism development in Cape Verde and have suggested greater participation of local communities. Social governance can be achieved by creating social value through community empowerment (Altinay, Sigala, & Waligo, 2016).

Improvements in residents' quality of life is accomplished through social governance. Research about the effect of tourism development on island residents' quality of life has been explored (Adanan, Radzi, Hanafiah, & Hamid, 2010; Ridderstaat, Croes, & Nijkamp, 2016). Chin and Hampton (2020) have suggested that quality of life is viewed as the satisfaction residents have with their day-to-day life. Jamal and Higham (2021) have pointed out the need for more consideration and ethical approaches to social justice in tourism including extending quality of life to a broader focus on the community well-being. In island tourism experiences the friendliness of residents may contribute to enjoyable tourism experiences. Cultivating positive host and tourist interactions will go a long way in realizing island destination branding and positioning goals. Social interactions among employees, residents, and tourists require a social governance framework that reinforces principles of diversity and equality. Based on a policy analysis of small island tourism, Giampiccoli, Mtapuri, and Dłużewska (2021) have argued for new paths for self-determination and social justice using a community-based tourism framework. Community-based tourism initiatives reverse the negative social impacts through an applied social governance framework of tourism awareness, education, and training (Giampiccoli et al., 2021). In addition, business relationships between members in the business community are important to support building sustainable partnerships that benefit destination development and prosperity of local communities.

Great social sustainability can be supported by broadening the social benefits of tourism. From a small island perspective of limited economic opportunities, residents seek employment opportunities in the tourism sector, therefore low salaries and wages paid to employees in the tourism and hospitality industries on the islands must be addressed (Baum, 2012). Low compensation rates affect not only lifestyle affordability but also access to health and education services that contribute to well-being. Zopiatis and Theocharous (2022) have explored career changes of hospitality managers in a small island and have found that anti-social working houses, work–family conflict, particularly after a change in marital status, becoming parents, and organizational aspects such as owner interference in the operations contribute to career changes. Work–life balance and healthy living experiences must be supported by a social policy framework that ensures people employed in the tourism sector obtain the same or greater benefits than the general population. Social governance actions must be clearly defined and articulated through policy instruments that are well-formulated by a consultation process with tourism stakeholders. Mitchell and Reid (2001) have attempted to socially integrate tourism using a planning tool that involves community power, awareness, and unity for island tourism in Peru. An understanding of the social benefits of island tourism should be identified and nurtured.

The social benefits of island tourism need to be highlighted. The creation of leisure and recreational spaces also benefits locals and residents. Such practices should be supported to build liveable communities with breathing green spaces for healthy living. The social conditions on islands can be considered in relation to the built-up infrastructure that supports social activities using a geo-informatic technique (Jungpanich & Waiyasusri, 2021). Sponsorship of social engagements, sporting activities, and events assists with bringing about social benefits in local communities. Social cohesion that supports tourism development can be cultivated through social governance policies (Lopes, Moreno Pires, & Costa, 2020). Entrepreneurial activities must be supported to create an innovative culture in the tourism sector. Entrepreneurs must be given opportunities to contribute to the tourism value chain in every aspect of the chain (Altinay et al., 2016). Boukas and Chourides (2016) have proposed that niche tourism products, such as ecotourism, provide social entrepreneurship opportunities in Cyprus that benefited local communities. The creation of a facilitating environment for local investment will build positive support for local tourism initiatives and build resident support for tourism investment (Boukas & Chourides, 2016). Islands have been known to seek foreign direct investment because of the financial risk that exists in small economies, however, greater support for partnerships between local and foreign investors as a general policy will go a long way to ensure local support for foreign investment (Barrowclough, 2007). Policies that ensure responsible foreign investment must be considered carefully to leave space for local entrepreneurs to contribute to tourism development in islands.

8.3 Island tourism governance environmental strategies

Environmental resources are a primary attraction on islands. Island destinations are challenged to balance the demands of a growing population and tourism industry. Kurniawan, Adrianto, Bengen, and Prasetyo (2019) have noted the increasing pressure of population growth and tourist arrivals on the environmental system on islands. In terms of the environmental management practices of residents, such practices are needed to improve the health, sanitation, and liveable island communities. Yao et al. (2023) have compared the stressful experiences of residents in three island communities, Oʻahu, Sãn Miguel, and Falmouth, and have noted traffic, crowding, congestion, and pollution among other stressors. To reduce stress, the standard of living in resort communities should be replicated in local communities through planned developments. Issues around unplanned development not only affect residents but can affect the aesthetics and attractiveness of the island scenery, particularly in urbanized small island environments (Nunkoo & Ramkissoon, 2010). The living conditions of employees in the hospitality industry are very important for good public health practices in the hospitality industry, since the environmental conditions can affect public health including food borne diseases. Apostolopoulos and Sönmez (2002) have assessed health risks for tourists in islands and have noted the need for mitigation planning and prevention. An outbreak of a disease or other public health issue is a matter for occupational health and safety teams to investigate the source and manage the problem. A food-poisoning incident on an island destination must be treated as a serious matter, particularly as there may be health facilities and services limitations, and must be resolved in the shortest possible time.

Environmental governance strategies in relation to waste management are critical for the sustainability of islands as tourist destinations. The expanse of the various types of waste, sewage, solid, food, becomes a difficult challenge for island authorities. Waste management is a major issue in island destinations (Graci, 2016; Kelman, 2019). Diaz-Farina, Díaz-Hernández, and Padrón-Fumero (2020), based on their study of the island of Tenerife, have noted that while tourism contributes to solid waste generation, fees are set using residential waste flows, and this causes budget imbalances. In some instances, the physical plants have not been developed to handle large capacities of waste. In such instances, the waste may have to be exported, dumped, or recycled and used in beneficial ways. In the context of waste generation, Telesford (2022) has proposed a model that restructures island tourism based on inflows and outflows. Specific to the issue of waste management, effective strategies are included in regulations and laws. In principle, waste management involves refuse, reduce, reuse, repurpose, and recycle (Kelman, 2019), and techniques such as composting and biogas generation should be explored. Anugrah (2022) with reference to Gili Trawangan, Indonesia, has suggested community-based waste management approach that involves waste transportation services, waste banks, recycling, and clean-ups. Technological advances in waste management must be researched, developed, and implemented. Waste reduction in the hospitality industry, including the consumption practices of both stayover and

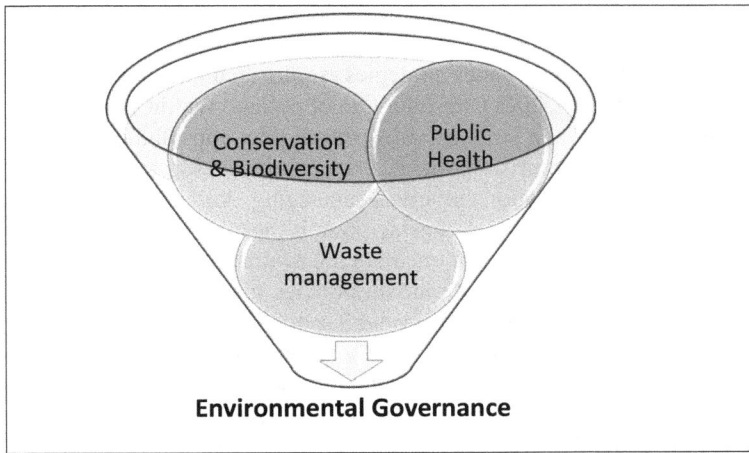

Environmental Governance

Figure 8.1 Environmental governance strategies.

cruise visitors, can go a long way in managing the tourists' consumption of goods on islands. The use of environmentally friendly products can help treat the breakdown of waste in a manner that would not harm the environment.

An island's ecosystem is comprised of both land and marine environmental resources, and land-based activities affect the state of the marine ecosystem. Conservation and preservation are two key concepts in the management of environmental resources. The biodiversity of island destinations must be maintained by appropriate controls that limit access to protected areas (Abreu, Tardieu, & Abreu, 2022; Carlsen, 2017). Small islands are particularly threatened by large influxes of tourists and without building awareness of protective measures and controls, irreversible damage and environmental loss are likely. An environmental audit, ecosystem changes, and resource management strategies are important for biodiversity management (Graci, 2016). Environmental governance involves specialized activities and island destinations may partner with agencies that have the necessary expertise to meet the peculiar needs of environmental governance on islands. A framework of environmental governance strategies has been proposed in Figure 8.1.

8.4 Island tourism governance political strategies

Political governance may be viewed as the formulation and implementation of policies that guide the development of a nation or territory (Fayos-Solà & Alvarez, 2014). Public policy is instituted through a political governance system called the public sector. The nuances of islands that were former colonies or still colonies affect the existing political governance strategies. Celata and Sanna (2012) have noted socio-political conflicts in the conservation of the Galápagos Islands and have suggested a post-political space to build consensus. Cultural practices may also affect political governance and sustainable development as one political party may dominate the political process (Yasarata, Altinay, Burns, & Okumus, 2010). While, tourism political governance strategies require an understanding of the political economy (Bramwell, 2011), the adoption of policies within an island tourism system should be based on a well-informed stakeholder consultation process. The result of well-formulated tourism policies is a tourism development process.

Tourism policy formulation involves managing knowledge (McLeod & McNaughton, 2015). An important consideration is the availability of tourism data to inform tourism policies. Information about the available data and the use of that data will further the formulation of tourism policies for an island tourism destination (McLeod, 2023a). Tourism policy-making ought not to be guess work. The range of policies that are needed to steer the tourism sector should be comprehensive, relevant, and timely to ensure effective policy implementation. Timely policies are essential to recover from external shocks in island destinations that face a range of vulnerabilities. Alejziak (2011) has argued for the broadening of the scope of tourism policies from a region to a global tourism policy given the internationalization of the tourism industry. Tourism growth and development must be guided by policies that will optimize benefits and minimize costs. The extent to which external factors may limit the realization of benefits has to be considered in effective policy formulation.

Policy implementation is perhaps the greatest challenge of political governance. The monitoring of the performance of policies is needed to inform whether the policy has achieved the planned result for island tourism development. Data are needed to adjust and make tourism policy changes and adapt to changing conditions in islands to build resilience (Bangwayo-Skeete & Skeete, 2021). Movono and Scheyvens (2022) have pointed out in the case of Fiji, how politics played a role in the actions taken during the pandemic. Both politics and policies are intertwined in island tourism governance. Tourism policies are political instruments, and the gamut of tourism policies may not be applicable across the board and therefore relevancy of tourism policies is important for its implementation within a certain context. Policy failure occurs (Becken & Loehr, 2022), and within the context of political governance the reasons for policy failure and the adjustments needed to reinstitute effective policies must be understood. A policy-making system is cyclical and adaptive, and sustainable governance means that transparency and accountability exist.

8.5 Island tourism governance economic strategies

Island governments use economic governance to facilitate the development of the tourism sector. Economic governance relates to how effectively decisions are made to manage an economy (Eagles, 2009). Economic governance strategies include micro-economic management and macro-economic management. Micro-economic management relates to the activities within the private sector to foster growth and development. Macro-economic management involves fiscal management such as taxation, monetary management such as increasing the money supply, and exchange rate policy. Exchange rates may fluctuate from time to time. Kuncoro (2016) has suggested that exchange rate policy directly affects the price of imported goods, that may be needed for the tourism industry and the cost of an island tourism experience. The amount of money a visitor has to spend, while on vacation, is directly related to exchange rate policy. Dropsy, Montet, and Poirine (2020) have suggested that tourist destination choice is related to the real exchange rate, a price ratio between the destination and origin countries.

Economic governance in island tourism requires greater focus on investment policies. Tourism activities in small island states facilitate economic growth, however, insufficient domestic capital has resulted in foreign direct investment (Roudi, Arasli, & Akadiri, 2019). Economic governance strategies may be legislated as a Hotel Development Act or Tourism Development Act. The purpose of these instruments is to create facilitating conditions to encourage and enable greater investment in tourism resources. The integration of micro- and macro-economic policies within a direct instrument for tourism development is a preferred strategy to enforce certain practices in the administration of the legislation. Alberts and Baldacchino (2017) have pointed out that in the case of the Caribbean a definite strength is the attraction of immigrant investors, and this strength has resulted in repeat visitors and time-share owners. Such economic governance strategies have resulted in contributing to economic resilience.

Increases in economic activities as a result of tourism, particularly in wholesale, retail, and business services are evident in islands (Pratt, 2015). Economic governance relates to business support for micro-, small- and medium-sized enterprises in an island's tourism sector. Granting small business loans, preferential interest rates, and longer repayment periods to facilitate financing for entrepreneurs in the tourism sector must be encouraged (Wright, 2021). Development of the digital capacity of these businesses to enable participation in every aspect of the tourism value chain is a key management strategy. Private sector stakeholders require a business environment that allows investment. Government policies should guide the areas for investment, and the necessary support infrastructure and financing can realize plans for tourism development. Increasing the money supply by lowering interest rates is a macro-economic policy that directly affects spending in the economy including investment in tourism. Saxena, Thaithong, and Tsagdis (2015) have noted the need for driving green economy growth through fiscal policies. Macro-economic policy as it relates to fiscal policies is important for government revenue collection, however, the private sector's ability to re-invest and create quality tourism products and services may be affected. A balance must be found to ensure that the tourism product develops in a sustainable way. If it means that government revenue can be directly reinvested in tourism products and services, then this must be done to ensure sustainability of tourism supply.

8.6 Conclusion

Tourism social, environmental, political, and economic governance strategies, for sustainable development of island tourism, are explored in this chapter. A framework for environmental governance strategies is detailed. Island tourism governance requires a refocusing from the dominant area of economic governance to handle the challenges of social and environmental governance. An island economy is cyclical with policy changes affecting all aspects of sustainable island development. By addressing governance strategies in a wholistic manner, the argument is that chances for sustainable development will be improved. Focusing on one area will lead to the detriment of the other areas and therefore a balanced approach is preferred.

Chapter 8 discussion questions

1 Explain the similarities and differences between social, environmental, political, and economic governance strategies.
2 Design and apply a framework for the four governance strategies outlined in the chapter.
3 Discuss the viewpoint the sustainable island development governance is dominated by environmental governance strategies.

References

Abreu, C., Tardieu, F., & Abreu, A. D. (2022). Conservation tourism in Pangatalan island, Palawan UNESCO Biosphere Reserve. In M. Novelli, J. M. Cheer, C. Dolezal, A. Jones, & C. Milano (Eds.), *Handbook of niche tourism* (pp. 38–48). Cheltenham, UK: Edward Elgar Publishing.

Adanan, A., Radzi, S. M., Hanafiah, M. H. M., & Hamid, Z. (2010). *Tourism development and its impacts to residents' quality of life: Case of Tioman Island.* Paper presented at the Proceedings of the 4th Tourism Outlook & 3rd ITSA conference, Shah Alam, Malaysia.

Alberts, A., & Baldacchino, G. (2017). Resilience and tourism in islands: Insights from the Caribbean. In R. Butler (Ed.), *Tourism and resilience* (pp. 150–162). Wallingford, UK: CABI.

Alejziak, W. (2011). A global tourism policy – utopia, alternative or necessity? *Folia Turistica, 25*(1), 313–356.

Altinay, L., Sigala, M., & Waligo, V. (2016). Social value creation through tourism enterprise. *Tourism Management, 54*(June), 404–417.

Anugrah, G. (2022). Interconnection between business and activism in managing waste: Case in community-based waste organization in small island tourism of Gili Trawangan, Indonesia. *Journal of Indonesian Tourism and Development Studies, 10*(2), 95–104.

Apostolopoulos, Y., & Sönmez, S. (2002). Disease mapping and risk assessment for public health and sustainable tourism development in insular regions. In Y. Apostolopoulos &

D. J. Gayle (Eds.), *Island tourism and sustainable development: Caribbean, Pacific and Mediterranean experiences* (pp. 225–248). Westport, CT: Praeger Publishers.

Bangwayo-Skeete, P. F., & Skeete, R. W. (2021). Modelling tourism resilience in small island states: A tale of two countries. *Tourism Geographies, 23*(3), 436–457.

Barrowclough, D. (2007). Foreign investment in tourism and small island developing states. *Tourism Economics, 13*(4), 615–638.

Baum, T. (2012). Human resource management in tourism: A small island perspective. *International Journal of Culture, Tourism and Hospitality Research, 6*(2), 124–132.

Becken, S., & Loehr, J. (2022). Tourism governance and enabling drivers for intensifying climate action. *Journal of Sustainable Tourism*, 1–19. doi:10.1080/09669582.2022.2032099

Bernardo, E., & Jorge, F. (2019). Are local residents able to contribute to tourism governance? – Impacts and perceptions in Cape Verde. *PASOS Revista de Turismo y Patrimonio Cultural, 17*(3), 611–624.

Bichler, B. F. (2021). Designing tourism governance: The role of local residents. *Journal of Destination Marketing & Management, 19*(March), 100389.

Boukas, N., & Chourides, P. (2016). Niche tourism in Cyprus: Conceptualising the importance of social entrepreneurship for the sustainable development of islands. *International Journal of Leisure and Tourism Marketing, 5*(1), 26–43.

Bramwell, B. (2011). Governance, the state and sustainable tourism: A political economy approach. *Journal of Sustainable Tourism, 19*(4–5), 459–477.

Carlsen, J. (2017). A systemic approach to tourism in island states. In R. N. Ghosh & M. A. B. Siddique (Eds.), *Tourism and economic development: Case studies from the Indian Ocean region* (pp. 94–103). London: Routledge.

Celata, F., & Sanna, V. S. (2012). The post-political ecology of protected areas: Nature, social justice and political conflicts in the Galápagos Islands. *Local Environment, 17*(9), 977–990.

Chin, W. L., & Hampton, M. (2020). The relationship between destination competitiveness and residents' quality of life: Lessons from Bali. *Tourism and Hospitality Management, 26*(2), 311–336.

Diaz-Farina, E., Díaz-Hernández, J. J., & Padrón-Fumero, N. (2020). The contribution of tourism to municipal solid waste generation: A mixed demand-supply approach on the island of Tenerife. *Waste Management, 102*(February), 587–597.

Dredge, D., & Jamal, T. (2013). Mobilities on the Gold Coast, Australia: Implications for destination governance and sustainable tourism. *Journal of Sustainable Tourism, 21*(4), 557–579.

Dropsy, V., Montet, C., & Poirine, B. (2020). Tourism, insularity, and remoteness: A gravity-based approach. *Tourism Economics, 26*(5), 792–808.

Eagles, P. F. (2009). Governance of recreation and tourism partnerships in parks and protected areas. *Journal of Sustainable Tourism, 17*(2), 231–248.

Fayos-Solà, E., & Alvarez, M. D. (2014). Tourism policy and governance for development. In E. Fayos-Solá, M. D. Alvarez, & C. Cooper (Eds.), *Tourism as an instrument for development: A theoretical and practical study* (pp. 101–124). Leeds: Emerald Group Publishing.

Giampiccoli, A., Mtapuri, O., & Dłużewska, A. (2021). Sustainable tourism and community-based tourism in small islands: A policy analysis. *Forum Geografic, 20*(1), 92–103. doi:10.5775/fg.2021.057.i

Graci, S. (2016). Collaboration and partnership development for sustainable tourism. In K. Meyer-Arendt & A. A. Lew (Eds.), *Understanding tropical coastal and island tourism development* (pp. 25–42). Abingdon: Routledge.

Holladay, P. J., & Powell, R. B. (2013). Resident perceptions of social–ecological resilience and the sustainability of community-based tourism development in the Commonwealth of Dominica. *Journal of Sustainable Tourism, 21*(8), 1188–1211.

Jamal, T., & Higham, J. (2021). Justice and ethics: Towards a new platform for tourism and sustainability. *Journal of Sustainable Tourism, 29*(2–3), 143–157.

Jungpanich, P., & Waiyasusri, K. (2021). Spatial assessment of built-up and recreation expansion using geo-informatic technique in Koh Chang Island, Thailand. *Geo Journal of Tourism and Geosites, 39*(4), 1501–1506.

Kelman, I. (2019). Critiques of island sustainability in tourism. *Tourism Geographies, 23*(3), 397–414.

Kinseng, R. A., Nasdian, F. T., Fatchiya, A., Mahmud, A., & Stanford, R. J. (2018). Marine-tourism development on a small island in Indonesia: Blessing or curse? *Asia Pacific Journal of Tourism Research, 23*(11), 1062–1072.

Kuncoro, H. (2016). Do tourist arrivals contribute to the stable exchange rate? Evidence from Indonesia. *Journal of Environmental Management and Tourism, 1*(13), 63–79.

Kurniawan, F., Adrianto, L., Bengen, D. G., & Prasetyo, L. B. (2019). The social–ecological status of small islands: An evaluation of island tourism destination management in Indonesia. *Tourism Management Perspectives, 31*(July), 136–144.

Lopes, V., Moreno Pires, S., & Costa, R. (2020). A strategy for a sustainable tourism development of the Greek Island of Chios. *Tourism: An International Interdisciplinary Journal, 68*(3), 243–260.

McLeod, M. (2023a). Managing Caribbean tourism data ecosystems. *Current Issues in Tourism*, 1–17. doi:10.1080/13683500.2023.2288665

McLeod, M. (2023b). Resilience building Caribbean Tourism. In G. Sinclair-Maragh (Ed.), *The dynamics of Caribbean tourism, opportunities, challenges and a re-imagined future* (pp. 1–21). Kingston: University of Technology, Jamaica Press.

McLeod, M., & McNaughton, M. (2015). *Knowledge-based tourism policy formulation, as an application of open data in Caribbean tourism.* Paper presented at the Proceedings of the International Conference on Tourism (ICOT2015), From Tourism Policy into Practice: Issues and Challenges in Engaging Policy Makers and End Users, London.

Mitchell, R. E., & Reid, D. G. (2001). Community integration: Island tourism in Peru. *Annals of Tourism Research, 28*(1), 113–139.

Movono, A., & Scheyvens, R. (2022). Tourism and politics: Responses to crises in Island states. *Tourism Planning & Development, 19*(1), 50–60.

Nunkoo, R., & Ramkissoon, H. (2010). Small island urban tourism: A residents' perspective. *Current Issues in Tourism, 13*(1), 37–60.

Parmawati, R., Pangestuti, E., Wike, W., & Hardyansah, R. (2020). Development and sustainable tourism strategies in Red Islands Beach, Banyuwangi Regency. *Journal of Indonesian Tourism and Development Studies, 8*(3), 174–180.

Pratt, S. (2015). The economic impact of tourism in SIDS. *Annals of Tourism Research, 52*(May), 148–160.

Rasoolimanesh, S. M., Ramakrishna, S., Hall, C. M., Esfandiar, K., & Seyfi, S. (2023). A systematic scoping review of sustainable tourism indicators in relation to the sustainable development goals. *Journal of Sustainable Tourism, 31*(7), 1497–1517. doi:10.1080/09669582.2020.1775621

Ridderstaat, J., Croes, R., & Nijkamp, P. (2016). The tourism development – Quality of life nexus in a small island destination. *Journal of Travel Research, 55*(1), 79–94.

Roudi, S., Arasli, H., & Akadiri, S. S. (2019). New insights into an old issue – Examining the influence of tourism on economic growth: Evidence from selected small island developing states. *Current Issues in Tourism, 22*(11), 1280–1300.

Said, F., & Farid, R. S. (2022). Marine tourism sustainability strategy with triple helix support: A case study of West Sulawesi Marine Tourism. *Journal of Humanities and Social Sciences Studies, 4*(4), 330–335.

Saxena, G., Thaithong, N., & Tsagdis, D. (2015). Investigating the transition of the tourism industry towards a green economy in Samui Island, Thailand. In M. V. Reddy & K. Wilkes (Eds.), *Tourism in the green economy* (pp. 310–324). Abingdon: Routledge.

Telesford, J. N. (2022). Restructuring island tourism: Using the socioeconomic metabolism (SEM) and multilevel perspective (MLP) as models for transitioning to sustainable island tourism. In I. Bethell-Bennett, S. Rolle, J. Minnis, & F. Okumus (Eds.), *Pandemics, disasters, sustainability, tourism* (pp. 109–123). Leeds, UK: Emerald Publishing.

Ternel, M., & Greyling, L. (2018). *An assessment of sustainable tourism and its opportunities in Mauritius.* Paper presented at the International Conference on Tourism Research, Kavala, Greece.

Wright, M. (2021). Funding proposals for new tourism ventures. In A. Spenceley (Ed.), *Handbook for sustainable tourism practitioners* (pp. 110–130). Cheltenham, UK: Edward Elgar Publishing.

Yao, X., Jordan, E. J., Spencer, D. M., Lesar, L., Vieira, J. C., Vogt, C. A., … Moran, C. (2023). A comparison of tourism-related stressors experienced by residents of three island destinations. *Tourism Geographies, 25*(4), 1251–1272.

Yasarata, M., Altinay, L., Burns, P., & Okumus, F. (2010). Politics and sustainable tourism development – Can they co-exist? Voices from North Cyprus. *Tourism Management, 31*(3), 345–356.

Zopiatis, A., & Theocharous, A. L. (2022). Career change in a small island tourism destination: Evidence from former hospitality managers. *International Journal of Hospitality & Tourism Administration, 23*(4), 834–859.

9 Tourism coordination and consensus building

9.1 Introduction

The propulsion of tourism development in island states must be supported by a common vision of the stakeholders for successful sustainable tourism development. The governance of the tourism sector is spread across various governmental, non-governmental, and private sector institutions that take decisive action to influence tourism in various ways. Kimbu and Ngoasong (2013) have argued that decentralization of tourism activities away from government has derailed a functional system of tourism policy actors. Such an action has resulted in competition and conflicts in power relationships, and therefore a centralization of a tourism policy network is recommended (Kimbu & Ngoasong, 2013). Keeping tourism development along a path that results in beneficial tourism must have the engagement of all stakeholders. Tourism consensus building involves obtaining an agreed best course of action for the success of the tourism industry. Such consensus is very difficult because of the range of interests and agendas of the diverse tourism stakeholders (Chia, Ramachandran, Ho, & Ng, 2018). Some of the challenges of tourism consensus building are both intrinsic and extrinsic to a tourism system. Intrinsic challenges are interactions of the supply and demand mechanisms, and the outcomes of those dynamics affect the decisions being made and the actions to be taken. Sustainable development requires combinations of decisions and actions at the local level to propel and change the performance of the tourism sector (Movono & Hughes, 2022). Extrinsic challenges relate to the macro environment, a lack of facilitating conditions and supporting mechanisms for tourism development and management.

Understanding consultation, conflict, coordination, and consensus-building processes can assist with integration of tourism stakeholder decisions and actions. Consultation involves obtaining feedback from tourism stakeholders and acting on that feedback in a way that stakeholders understand the reasons for any course of action. Conflicts occur as interests clash, and power struggles occur as those more powerful stakeholders gain ground in progressing plans for preferred tourism activities. Understanding the contributions of trade associations is important (McKercher, 2022). The role of trade associations to achieve the interests of

DOI: 10.4324/9781003435112-11

members is a political process making power and politics intricately involved in tourism development. To support tourism development, coordination seems to be middle ground between conflict and consultation with a working together of stakeholders to achieve a common interest.

9.2 Consultation processes

A consultation involves obtaining the views, ideas, preferences, and knowledge of tourism stakeholders to garner support and understanding for a particular course of action. A consultation does not necessarily result in agreement, but it will allow a bringing together of minds to thoroughly discuss an issue. For a consultation process to be successful, stakeholders who are most knowledgeable about the subject matter should be involved. Pathak, van Beynen, Akiwumi, and Lindeman (2022) have noted the importance of consultation workshops for climate adaptation planning in The Bahamas. In island states, local involvement in consultation processes can be constrained. In a post-slavery island, a lack of local engagement in tourism development may be related to educational constraints, lack of civicness, low initiative and self-confidence, and low risk-taking behaviour (Nicely & Sydnor, 2015). Consultation success also depends on unpacking the nature of the problem, and discussing issues that may cause or are related to the problem. An understanding of the purpose and goals to be achieved during the consultation process will assist with keeping the process on track.

Consultations are required for activities such as tourism planning and impact assessments. The challenge with the consultation process is finding a way to deliver on the recommendations or outputs of the consultation. Stakeholders involved in the consultation process have an expectation that their ideas will be considered and taken on-board. Time and a great deal of effort are invested in consultation processes, and therefore the information gathered must be meaningful to resolve the issue or problem at hand. Consultations must be structured in a way that the varying perspectives of stakeholders are being heard and accounted for. Conflict often arises without consultation from interest groups, and sustainable development becomes problematic with multiple interpretations of findings (Connell, 2018). Movono and Scheyvens (2022) have noted the Fiji government's designation of local indigenous coastal property as public property without consultation with the local owners. Such practices result in a loss of indigenous rights and access to resources (Movono & Scheyvens, 2022). Simão and Môsso (2013) have noted the practice of not including residents in consultation and decision-making processes for island tourism development. Movono and Hughes (2022) have recommended that consultation forms the basis of a partnership and relationship building to achieve the Sustainable Development Goals. Walker, Lee, and Li (2021) have found in Antigua and Barbuda, consultations enabled alliances and partnerships, including opportunities for exports and contributing towards sustainable tourism development.

A consultation forum must be open in a way that stakeholders feel comfortable airing their views. If an open forum is not appropriate for airing views, then interviews of key informants should be included. Balancing the interests of tourism stakeholders in taking certain ideas forward is challenging. Connell (2018) has noted that varying interests in island tourism development require a balance between development and sustainability. Grilli, Tyllianakis, Luisetti, Ferrini, and Turner (2021) have pointed out the trade-offs between tourism services and socio-cultural identities in small island developing states. A way forward must be found that can be mutually beneficial to residents of an island destination. Groups that seek to take advantage of the consultation opportunity to raise issues not relating to the resolution or issue should be notified of this to ensure that the consultation process remains on track to achieve a particular purpose. A consultation process must generate a wide range of possible solutions to the issue or problem. In addtion, consultation between tourism institutions is needed to devise effective tourism development solutions (Jordan, 2004). Solution ideation requires skill to empower stakeholders in solving some very difficult sustainable development problems. If there are disagreements during the consultation process, these should not be avoided but managed in a way to bring about understanding of the varying perspectives.

9.3 Conflict resolution strategies

Conflict resolution may be viewed as a soft skill to breakdown and analyse an issue in a way that a resolution is possible. Conflicts occur because of several reasons. Elliott (2020) has noted a nexus between politics and power articulated through a political environment of ideology, conflicts, policies, and priorities. Various tourism stakeholders are involved in boards, committees, and associations that contribute to the decision-making in the tourism sector (McLeod, Chambers, & Airey, 2018). Tourism stakeholders may have competing aims and objectives, and choices must be made as to the best way forward. In the tourism sector, conflicts may emerge from various sources. Sources of conflicts include government implementing a new policy or regulation, a business expanding its operations, tourism development in a local community, a local community's lack of access to free resources such as a beach, a lack of local opportunities for tourism businesses (Chia et al., 2018). The list of possible conflicts in the tourism sector is extensive. The first step in understanding a conflict is identifying the source of the conflict and detailing the scope and impact of the conflict. Conflicts may also occur because of challenges in the internal tourism economy, however, the external tourism forces may also be a source of conflict.

Power conflicts are evident in island tourism. Farmaki, Altinay, Botterill, and Hilke (2015), while exploring politics and sustainable tourism development in Cyprus, have noted the power struggles, a source of conflict, that occurred between global, national, and local tourism stakeholders. Church (2004) has pointed to the broader economic and political processes such as globalization, state restructuring, changes to civil society, and the changing power structures that shape tourism development. Nunkoo (2017) has noted the relational aspect of power in sustainable tourism and has argued that power differences exist between tourism stakeholders. Power through relationships requires exploration in tourism since power can steer tourism activities in a particular direction. Farmaki et al. (2015) have sought to explore the political dimension of sustainable tourism and have pointed to the external conduits of power that influence tourism development. Power is derived from the connections of various tourism stakeholders and the power each actor has may be utilized to contribute to the policy formulation for the tourism sector. Additionally, enclaves produce power inequalities, leakages, and are unsustainable (Saarinen & Wall-Reinius, 2019).

A conflict resolution is possible if the aspects of the conflict are within the control of the resolver. One way of resolving a conflict is to formalize the resolution in an instrument. In the context of sustainable development, such instruments must be carefully selected. While a policy may guide, a legislative instrument will be enforced. A destination may resort to such a mechanism to gain power to promote a tourism destination in conflict (Causevic & Lynch, 2013). In some instances, policies and legislation may give preferential treatment of large-scale tourism development at the detriment of the environment (Hampton & Jeyacheya, 2015). An absence of legislative power and regulations results in informal tourism developments that also affect the environment (Farmaki et al., 2015). Power applied through legislation has its shortcomings in terms of the selection of those appropriate policies to be legislated and in terms of enforcement of legislative instruments. Other instruments of power are needed to achieve political purpose in islands that are resource constrained. On one hand, if the resolution involves increasing resources, then the conflict cannot be resolved at that point in time. On the other hand, if the conflict arises because different groups lack understanding of their responsibilities in the tourism development process, then such a conflict can be resolved by clarification and agreement among the groups.

On a destination level, taking responsibility to resolve a conflict is not a simple activity. First, responsibility is assigned based on the source of the conflict. Second, the conflict must be thoroughly understood. Third, in some instances the entity seeking to resolve the conflict such as a government authority may be part of the conflict. Fourth, no one wants the conflict to become a greater problem than it is already and therefore an attempt to resolve may be delayed but to the detriment of those involved. Accountability must also be involved in conflict resolution (Curcija, Breakey, & Driml, 2019). The conflict resolution process itself should do no harm to the stakeholders involved. The roles of those involved in the conflict resolution process must be identified and clarified with those involved in the conflict resolution process. Reporting on the progress of conflict resolution is just as important as the process itself. A resolution should be within reach, and the matter should be on its way to no longer being a problem or issue. The mini case study about stopping the construction of an airport in Barbuda allows for the exploration of a conflict that affects access to the island.

Mini Case Study 7: Barbuda's halt from bouncing back after a hurricane

The country of Antigua and Barbuda depends on tourism. Revenues as a percentage of exports were over 80% over the period of 2016–2019 (Figure 9.1) (World Bank, 2023). The main island Antigua makes a claim of a beach for each day of the year. St. Johns, the capital of the country, is a picturesque town. While there is tourism growth in the country, the island of Barbuda faced the challenge of recovering from Hurricane Irma that damaged almost the entire island. The island of Barbuda was evacuated in 2017 because of the impacts of a hurricane. Land rights on the island give local Barbudans sole jurisdiction for the development of the island. As a result, the airport construction was challenged (Island Innovation, 2023). Two Barbudans filed a case to stop the construction of the airport because of its impact on the island's ecosystem (Taylor, 2018). The court rejected the appeal against the Development Control Authority's construction of an airport on Barbuda (Judicial Committee of the Privy Council, 2023). Given that an airport is a critical infrastructure that provides access to the island, a resolution of this issue had to be found in the shortest possible time.

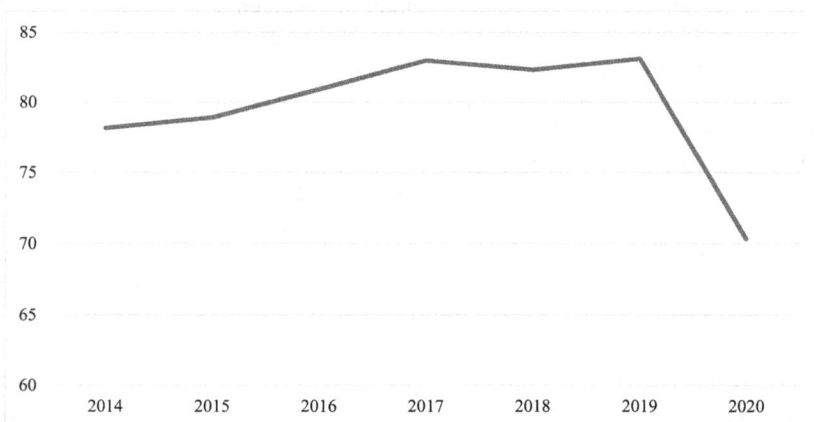

Figure 9.1 Antigua and Barbuda international tourism receipts as a percentage of total exports.

Source: World Bank (2023).

Mini Case Study 8 discussion questions

1 Review the circumstances about the construction of an airport in Barbuda and set out the consultation process that could have been engaged.
2 Consider the twin island context of Antigua and Barbuda and discuss the collaborative workings of institutions of island tourism development.
3 Explain recommendations that may set out a way forward for a beneficial outcome for the residents of Barbuda.

9.4 Coordination mechanisms

Coordination is perhaps the most difficult activity in the tourism sector. Tourism policy actors are interconnected through various tourism organizations, businesses, and institutions and represent those bodies in the management and development of tourism. Poor coordination of stakeholder activities in the implementation of sustainable tourism projects limits development opportunities in islands (Uysal & Modica, 2016). Tourism policy formulation and governance involve a system of policy actors, responsible for influencing tourism development, and government actors, private sector actors, and others are comprised in that system (Dredge, 2006; McLeod, 2023; Pforr, 2006; Wray, 2009). Tourism governance progress is built on ensuring that the benefit from tourism outweighs the costs. Understanding tourism governance is complicated by the myriad of cross-sectional, cross-sectoral policies that contribute to tourism development and the multiplicity of roles played by various tourism actors. Albrecht (2010) has noted in the case of Stewart Island, New Zealand, the lack of coordination and support in implementation of tourism strategy, as stakeholder action may be limited to community involvement. The power of policy actors at the community level must be determined, as power is important to achieve political purpose, and power is important for community involvement.

The tourism product and/or service is arrayed with diverse, wide-ranging elements, often across international borders, and for island destinations, coordination activities are complicated. This makes coordination an activity that must be carefully mapped out. A coordinating mechanism may be viewed as the connectivity between and among entities involved in meeting a common goal or purpose. Coordination means that the entities are working together in a harmonious relationship that will achieve the desired outcomes. Charlie, King, and Pearlman (2013) have suggested that in the environmental governance of small islands collaboration occurs by coordination relationships between tourism stakeholders. For coordination to work, all stakeholders must buy into the stated purpose of the coordination. Klint et al. (2012) have supported the collaborative approach for effective climate adaptation policy implementation in the Pacific Islands. If there are conflicts, such as when a policy changes, then conflict resolution is needed before coordination can occur. For this reason, the nature of governance needs to be understood from a network perspective that illustrates the connections between actors and among several tourism stakeholders (Kimbu & Ngoasong, 2013). The resource connections between actors are just as important as their position within a network structure as those actors that are better connected can benefit from network resources and influence network outcomes (Borgatti & Halgin, 2011).

In the tourism sector, government is often seen as the main coordination actor. Government coordination can be viewed as a formal arrangement. Saxena (2005) has noted that formal networks take on hierarchical arrangements, and informal networks are open-ended. For sustainable tourism development to occur, government action must be supported by the gamut of tourism industry stakeholders and therefore coordination relationships may be varied. Coordination involves matrix relationships formed based on purposeful actions to achieve a result and these relationships are comprised of both formal and informal relationships. Government agencies involved in managing and developing the tourism sector report to a minister who has responsibility for tourism. Although responsibility is assigned to a minister, control of tourism activities is another matter. In the public sector, integration of activities to avoid duplication of efforts in rolling out tourism activities is a critical necessity. For coordination in sustainable development to be effective, those tourism stakeholders must have a common interest that aligns with government tourism policies and actions. With alignment comes cooperation so that the coordinating actions work smoothly. Cooperation of tourism industry stakeholders must be voluntary rather than coerced through penalties and restrictions.

Clarification of the coordination structure also helps for better understanding of the activities being undertaken to steer the tourism and hospitality industries along a sustainable development path. A coordination structure should be evolving ensuring that the competencies and capabilities of those involved are being deployed effectively. Williams, You, and Joshua (2020) have examined small business resilience in St. Helena, a small remote island, and have found that knowledge-sharing, long-term relationships, supportive coordination, and trust were the most important aspects in building resilience. Supportive collaborative structures on islands involve connections and communication during the coordination of tourism activities. A clear communication plan including crisis communication strategy is essential for coordination of the tourism sector on islands. Informing tourism stakeholders of plans, changes, schedules, and resources will assist in the achievement of desired goals and outcomes. Resource wastage on islands with resource constraints must be avoided. Timothy and Tosun (2003) have pointed to the need for public agencies to coordinate efforts and reduce wastage of overlapping projects. Coordination involves prioritization based on set goals and objectives.

9.5 Consensus-building strategies

If all else fails, consultation, conflict resolution, and coordination, consensus building may be the only viable strategy for effective tourism governance in island destinations. Consensus building in tourism may be viewed as an unspoken agreement of the way forward for tourism development. It is an activity to achieve a balancing of stakeholder interests. Some may view a consensus as a win-win or mutual benefit. Consensus building may occur during the other governance networking activities (Partelow & Nelson, 2020). Clarity on the causes of a problem or issue is needed and a range of various solutions that stakeholders are likely to agree with should be determined. As part of a consensus-building strategy, various best practices, data and information, and research must be obtained to gain support for possible solutions.

Consensus building takes time and stakeholders must be willing to participate fully in the process and be committed to the solutions for the process to be effective (Bichler, 2021; Roxas, Rivera, & Gutierrez, 2020). Consensus building is a critical activity to face sustainability challenges (Roxas et al., 2020). Mitchell and Reid (2001) have pointed out that consensus building is a painstaking activity within the complexity of tourism planning for island development. Tourism policies require the involvement of public, private, and voluntary sectors (Fayos-Solá, 1996). While planning approaches have advantages, the implementation of sustainable tourism development plans has been a major downfall, as without obtaining broader stakeholder consensus, cooperation may not occur (Fayos-Solá, 1996; Tosun, 2001).

Clear policies, which stakeholders understand, must be articulated to create consensus for sustainable development. For example, Sheller (2021) has pointed to the reconstruction of Caribbean tourism through a regenerative approach, including the protection of water resources, coastal areas, and forests, as current forms of tourism have placed a strain on the ecology of pristine marine and land environments. Under such circumstances, sustainable development policy directions have been articulated, but in the Caribbean the issue remains as to the stakeholder involvement and control (Baker, 2022). The notion that Caribbean tourism is dominated by foreign investment remains (Spencer et al., 2023). While consensus building does not need to be a democratic process, the workings of power in tourism development require elaboration. Consensus building must be able to result in a likelihood that stakeholders will support the solutions. Consensus building must also involve a follow-up activity after solutions have been implemented to monitor solution effectiveness.

Consensus building is a collaborative process with stakeholders having mutual respect. In small island destinations with homogeneous populations and close relationships, consensus building may occur more readily than in larger islands with fragmented relationships and limited social support. While heterogeneity will bring about richer and diverse solutions through innovation, diversity may result in disagreements and therefore consensus building must be implemented based on the context of the island. Benefits of consensus building include cooperative behaviour, achievement of results, increased awareness of practices, and information sharing to improve processes. In the context of an archipelago with multiple islands balancing each island's interest adds to the complexity of consensus building (Bardolet & Sheldon, 2008).

9.6 Conclusion

Tourism governance requires strategies that will bring about effective results in the management and development of island tourism. Consultation, conflict resolution, coordination, and consensus building are approaches that bring together tourism stakeholders around a common purpose, sustainable development. Good island tourism governance is both the policy content and about the governance process. Improvement in governance processes requires collaborative strategies including coordination and consensus building as softer approaches to steer tourism along a sustainable development path. Sustainable development requires a consensus that certain courses of actions must be taken. Bringing about sustainability is not a given but an understanding of why and how sustainability is the best way forward. By imploring tourism stakeholders to build consensus for sustainability, island destinations will enable a sustainable development process.

Chapter 9 discussion questions

1 Consider the ways tourism governance is coordinated in an island destination and design a framework of the activities.
2 Distinguish between soft and hard measures of tourism governance in island destinations.
3 Discuss coordination differences of the tourism sector in large and small island destinations.

References

Albrecht, J. N. (2010). Challenges in tourism strategy implementation in peripheral destinations—The case of Stewart Island, New Zealand. *Tourism and Hospitality Planning & Development, 7*(2), 91–110.

Baker, D. M. A. (2022). Caribbean tourism development, sustainability, and impacts. In C. Cannonier & M. Galloway Burke (Eds.), *Contemporary issues within Caribbean economies* (pp. 235–264). Cham: Palgrave Macmillan.

Bardolet, E., & Sheldon, P. J. (2008). Tourism in archipelagos: Hawai'i and the Balearics. *Annals of Tourism Research, 35*(4), 900–923.

Bichler, B. F. (2021). Designing tourism governance: The role of local residents. *Journal of Destination Marketing & Management, 19*(March), 100389.

Borgatti, S. P., & Halgin, D. S. (2011). On network theory. *Organization Science, 22*(5), 1168–1181.

Causevic, S., & Lynch, P. (2013). Political (in)stability and its influence on tourism development. *Tourism Management, 34*(February), 145–157.

Charlie, C., King, B., & Pearlman, M. (2013). The application of environmental governance networks in small island destinations: Evidence from Indonesia and the Coral Triangle. *Tourism Planning & Development, 10*(1), 17–31.

Chia, K.-W., Ramachandran, S., Ho, J.-A., & Ng, S. S.-I. (2018). Conflicts to consensus: Stakeholder perspectives of Tioman Island tourism sustainability. *International Journal of Business & Society, 19*(1), 159–174.

Church, A. (2004). Local and regional tourism policy and power. In A. A. Lew, C. M. Hall, & A. M. Williams (Eds.), *A companion to tourism* (pp. 555–568). Malden, MA: Blackwell.

Connell, J. (2018). Islands: Balancing development and sustainability? *Environmental Conservation, 45*(2), 111–124.

Curcija, M., Breakey, N., & Driml, S. (2019). Development of a conflict management model as a tool for improved project outcomes in community based tourism. *Tourism Management, 70*(February), 341–354.

Dredge, D. (2006). Policy networks and the local organisation of tourism. *Tourism Management, 27*(2), 269–280.

Elliott, J. (2020). *Tourism: Politics and public sector management.* Abingdon, UK: Routledge.

Farmaki, A., Altinay, L., Botterill, D., & Hilke, S. (2015). Politics and sustainable tourism: The case of Cyprus. *Tourism Management, 47*(April), 178–190.

Fayos-Solá, E. (1996). Tourism policy: A midsummer night's dream? *Tourism Management, 17*(6), 405–412.

Grilli, G., Tyllianakis, E., Luisetti, T., Ferrini, S., & Turner, R. K. (2021). Prospective tourist preferences for sustainable tourism development in small island developing states. *Tourism Management, 82*(February), 104178.

Hampton, M. P., & Jeyacheya, J. (2015). Power, ownership and tourism in small islands: Evidence from Indonesia. *World Development, 70*(June), 481–495.

Island Innovation (2023). 'Billionaire club': The tiny island of Barbuda braces for decision on land rights and nature. Retrieved from https://islandinnovation.co/news/billionaire-club-the-tiny-island-of-barbuda-braces-for-decision-on-land-rights-and-nature/

Jordan, L.-A. (2004). Institutional arrangements for tourism in small twin-island states of the Caribbean. In D. T. Duval (Ed.), *Tourism in the Caribbean: Trends, development, prospects* (Vol. 3, pp. 99–118). London: Routledge.

Judicial Committee of the Privy Council (2023). John Mussington and another (Appellants) v development control authority and 2 others (Respondents) (Antigua and Barbuda). Retrieved from www.jcpc.uk/cases/jcpc-2021-0116.html

Kimbu, A., & Ngoasong, M. (2013). Centralised decentralisation of tourism development: A network perspective. *Annals of Tourism Research, 40*(January), 235–259. https://doi.org/10.1016/j.annals.2012.09.005

Klint, L. M., Wong, E., Jiang, M., Delacy, T., Harrison, D., & Dominey-Howes, D. (2012). Climate change adaptation in the Pacific Island tourism sector: Analysing the policy environment in Vanuatu. *Current Issues in Tourism, 15*(3), 247–274.

McKercher, B. (2022). The politics of tourism: The unsung role of trade associations in tourism policymaking. *Tourism Management, 90*(June), 104475.

McLeod, M. (2023). Tourism policy networks in four Caribbean countries. *Annals of Tourism Research Empirical Insights, 4*(2), 100113.

McLeod, M., Chambers, D., & Airey, D. (2018). A comparative analysis of tourism policy networks. In M. McLeod & R. Croes (Eds.), *Tourism management in warm-water island destinations* (pp. 77–94). Wallingford, UK: CABI.

Mitchell, R. E., & Reid, D. G. (2001). Community integration: Island tourism in Peru. *Annals of Tourism Research, 28*(1), 113–139.

Movono, A., & Hughes, E. (2022). Tourism partnerships: Localizing the SDG agenda in Fiji. *Journal of Sustainable Tourism, 30*(10), 2318–2332.

Movono, A., & Scheyvens, R. (2022). Tourism and politics: Responses to crises in Island states. *Tourism Planning & Development, 19*(1), 50–60.

Nicely, A., & Sydnor, S. (2015). Rural tourism development: Tackling a culture of local nonparticipation in a postslavery society. *Journal of Travel Research, 54*(6), 717–729.

Nunkoo, R. (2017). Governance and sustainable tourism: What is the role of trust, power and social capital? *Journal of Destination Marketing & Management, 6*(4), 277–285.

Partelow, S., & Nelson, K. (2020). Social networks, collective action and the evolution of governance for sustainable tourism on the Gili Islands, Indonesia. *Marine Policy, 112*(February). doi:10.1016/j.marpol.2018.08.004

Pathak, A., van Beynen, P. E., Akiwumi, F. A., & Lindeman, K. C. (2022). Climate change in the strategic tourism planning for small islands: Key policy actors' perspectives from The Bahamas. In I. Bethell-Bennett, S. Rolle, J. Minnis, & F. Okumus (Eds.), *Pandemics, disasters, sustainability, tourism* (pp. 125–143). Leeds, UK: Emerald Publishing.

Pforr, C. (2006). Tourism policy in the making: An Australian network study. *Annals of Tourism Research, 33*(1), 87–108.

Roxas, F. M. Y., Rivera, J. P. R., & Gutierrez, E. L. M. (2020). Mapping stakeholders' roles in governing sustainable tourism destinations. *Journal of Hospitality and Tourism Management, 45*(December), 387–398.

Saarinen, J., & Wall-Reinius, S. (2019). Enclaves in tourism: Producing and governing exclusive spaces for tourism. *Tourism Geographies, 21*(5), 739–748.

Saxena, G. (2005). Relationships, networks and the learning regions: Case evidence from the Peak District National Park. *Tourism Management, 26*(2), 277–289.

Sheller, M. (2021). Reconstructing tourism in the Caribbean: Connecting pandemic recovery, climate resilience and sustainable tourism through mobility justice. *Journal of Sustainable Tourism, 29*(9), 1436–1449.

Simão, J., & Môsso, A. (2013). Residents' perceptions towards tourism development: The case of Sal Island. *International Journal of Development Issues, 12*(2), 140–157.

Spencer, A. J., Lewis-Cameron, A., Roberts, S., Walker, T. B., Watson, B., & McBean, L. M. (2023). Post-independence challenges for Caribbean tourism development: A solution-driven approach through Agenda 2030. *Tourism Review, 78*(2), 580–613.

Taylor, D. (2018). Work on Caribbean island airport halted by court ruling. Retrieved from www.theguardian.com/world/2018/aug/02/plans-for-airport-on-caribbean-island-of-barbuda-face-legal-challenge

Timothy, D. J., & Tosun, C. (2003). Appropriate planning for tourism in destination communities: Participation, incremental growth and collaboration. In S. Singh, D. J. Timothy, & R. K. Dowling (Eds.), *Tourism in destination communities* (pp. 181–204). Wallingford, UK: CABI.

Tosun, C. (2001). Challenges of sustainable tourism development in the developing world: The case of Turkey. *Tourism Management, 22*(3), 289–303.

Uysal, M., & Modica, P. (2016). Island tourism: Challenges and future research directions. In P. Modica & M. Uysal (Eds.), *Sustainable island tourism: Competitiveness and quality of life* (pp. 173–188). Wallingford, UK: CABI.

Walker, T. B., Lee, T. J., & Li, X. (2021). Sustainable development for small island tourism: Developing slow tourism in the Caribbean. *Journal of Travel & Tourism Marketing, 38*(1), 1–15.

Williams, C., You, J. J., & Joshua, K. (2020). Small-business resilience in a remote tourist destination: Exploring close relationship capabilities on the island of St Helena. *Journal of Sustainable Tourism, 28*(7), 937–955.

World Bank (2023). World development indicators data bank. Retrieved from https://datab ank.worldbank.org/source/world-development-indicators

Wray, M. (2009). Policy communities, networks and issue cycles in tourism destination systems. *Journal of Sustainable Tourism, 17*(6), 673–690.

Part III

Island tourism planning for sustainable development

10 Tourism planning frameworks

10.1 Introduction

Tourism planning in island destinations is easier said than done. Planning is an activity engaged with bringing about some destination change. Beyond a requirement for a development loan, a tourism plan must consider the sustainability of tourism activities within an island destination. For example, the Jamaica Hotel Act of 1890 provides for capital loans at 3% interest and duty-free allowance for construction materials and furniture (Spencer et al., 2023). With such incentives, hotel room growth has burgeoned, however, the sustainability of the island's tourism sector requires planning. This chapter focuses on the various types of tourism planning methods for sustainable development to provide alternative planning approaches for island destinations. While development and spatial planning activities seem popular approaches, scenario planning and outcome mapping methodology are least developed as tourism planning methods. Strategic planning has been included and outcome mapping methodology has been introduced as an approach to realize planning objectives.

One of the challenges of tourism planning is the involvement of key stakeholders, including the broader sectors in the economy, for wholistic planning idea generation. While tourism stakeholders are the first point of contact, the stakeholders indirectly related to the tourism sector may be left out to the detriment of planning outcomes. Maintenance of partnerships is particularly important for the sustainable development of islands (Movono & Hughes, 2022). Tourism is a multi-stakeholder integrated activity in island destinations. Firdaus and Endah (2015) have noted in Bangka Island the lack coordination among stakeholders and lack of communication between the tourism and mining sectors, and a limited role of communities in environmental conservation. In island economies, with tourism as a dominant industry, nearly everyone is involved, or in some way affected by tourism development directly or indirectly. Tourism planning on islands must be inclusive of all stakeholders and aspects of island life to ensure that the eventual tourism plan is implementable.

From a practical perspective, a tourism plan is not meant to be an expensive piece of literature sitting on someone's shelf and gathering dust (King, McVey, & Simmons, 2000), but a document that can create value in an island economy.

DOI: 10.4324/9781003435112-13

A lack of funding for the implementation of a tourism plan can derail the potential value of planning activities. Timing of the development and the implementation of tourism planning is also important. Also, stakeholders sometimes do not participate in tourism planning activities when results are not evident (Siti-Nabiha & Saad, 2015). With funding and appropriate timing, tourism plans can be realized. Another important aspect to consider is the master plan (Teniwut, Hamid, & Makailipessy, 2022). Tourism master plans are grand ideas, and without proper consideration of the dynamics of the tourism system, such plans can be unrealistic. The idea that all tourism development activity can be designed and mapped out in one master plan is a good one but often the lack of implementation means another planning activity should be embarked upon. New planning methodologies are therefore needed to improve chances of sustainable development.

10.2 Development planning

Development planning is a process in which the main objective of the plan is to engage in a step-by-step development process to realize an aim and accompanying objectives of the plan. Development planning achieves a stated purpose. A development plan may be viewed as a project to develop an area in keeping with the policies for the development. Development planning for tourism must consider the overall development strategy for the island. Croes (2022) has noted a delayed tourism development strategy in Curacao based on a belief that diversification will result in prosperity for the island. Lopes, Moreno Pires, and Costa (2020) have pointed out the need for strategies for sustainable tourism development in island destinations and have adopted the 17 Sustainable Development Goals as a planning framework for the Island of Chios. To evaluate tourism development, tourism plans often set out specific policy goals and objectives. The specific content of tourism policies is key to the development of a successful tourism sector (Caribbean Tourism Organization, 2020).

Development planning in an island context must be built on an exploration of core principles and practices (McLeod, Dodds, & Butler, 2021). The level at which planning is to be done should be determined and must have the support of key stakeholders in the local communities. A national sustainable development plan requires resourcing to a larger extent than a regional or local plan. Ponte, Couto, Pimentel, Sousa, and Oliveira (2021) have argued for planning at the local level to realize benefits for local communities. A commitment by government, business community, and the public at large, to the outcomes of a planning process engender appropriate planning outcomes. Canavan (2014) has noted the politics of allowing wealthy immigrants to build without planning permission and such actions have made the island less attractive. Planning for sustainable development also means integration of the political as well as economic, socio-cultural, and environmental elements of an island destination. A workable plan should set out realistic goals that are implementable within a certain timeframe. In that regard, it has been suggested that islands first evaluate available resources and then find a suitable market that fits the island's appeal (Azzopardi & Nash, 2016).

Development planning must also consider building a resiliency framework into the plans. Alberts and Baldacchino (2017) have pointed out the need for migration acceptance to support labour shortages in Caribbean islands as a necessity for resilience. The extent to which this can be done may be limited, as development planning may be limited in understanding forecasts and outlook of the island destination's disasters and crises. Pyke et al. (2021) have proposed an assessment of resilience factors and vulnerabilities and have pointed out a lack of adaptation as a constraint. Consideration of mitigation and adaptation strategies in the development of infrastructure and the built environment are important for the long-term success of an implemented development plan. One of the greatest challenges facing island destinations is the climatic forces that have resulted in rising sea levels and loss of marine environments. Development plans must adequately consider climatic forces to build resiliency and ensure sustainable development.

10.3 Scenario and strategic planning

Scenario planning has the parameters to build resiliency into sustainable development plans for island destinations. A scenario is a prediction of an occurrence, and scenario planning is a process of developing a plan based on the occurrence. Seyitoğlu and Costa (2022) have categorized the usefulness of scenario planning in tourism for understanding internal complexity and external uncertainties. The core activity in the scenario planning process is the development of the scenario. For island destinations, possible future scenarios include natural and economic disasters based on the physical and human geographies of islands. Pizzitutti et al. (2017) have built three population growth scenarios for understanding the interactions of physical and human systems in the Galápagos Islands. Scenario planning may also involve developing more than one scenario, possibly as many as four scenarios, and planning development based on the four scenarios. Page, Yeoman, Connell, and Greenwood (2010) have constructed two scenarios as a planning tool for uncertainty, one that involved urban growth and decline of rural and island tourism, and another based on tourism development within the constraints of transport infrastructure. Based on the scenarios, policies can be developed to address tourism planning needs (Page et al., 2010).

Long-range planning may be done using scenarios. While development plans are short to medium term, long-range planning eventually realizes the main aim of the development by considering adjustments being made during the implementation phase. McLennan, Pham, Ruhanen, Ritchie, and Moyle (2012) have supported long-range planning. Long-range planning is appropriate for sustainable development as the challenges that require plan adjustment and policy development are considered during the planning process. Mai and Smith (2018) have suggested that historical data are poor indicators of future events because of feedback loop shifts in the long term. Nonetheless, in building out scenarios and long-range planning, historical data can assist with the construction of the actual scenarios. Data and information are crucial, and artificial intelligence tools for planning are another consideration. The lack of data and information is perhaps the greatest hindrance in planning accurately for likely outcomes. The sources of information from a national statistics agency, international agency, or from a market research agency are important to facilitate the planning process. Agencies such as central banks and immigration departments have data that can readily inform data gaps about performance and resource needs for the tourism sector (McLeod, 2023). Equally important is being able to determine the data quality of the information being used. Data quality issues may occur based on errors in data collection and errors in data analysis. Plans made on inaccurate results from data will result in development failure and loss of an opportunity to utilize resources elsewhere.

Strategic planning should be considered as development of a plan based on a situational analysis of needed changes. The achievement of an island destination's strategic fit with the dynamics of the tourism system is very important to realize sustainable development. Roxas, Rivera, and Gutierrez (2020) have noted while policies are important for growing tourism activities, strategic planning is necessary for environmental sustainability, and to ensure benefits from tourism are realized in local communities. Strategic planning activities are to be informed by stakeholders and tourism experts through conducting a gap analysis, and resource planning activities to implement the strategic plan are also important. Lopes et al. (2020) have noted that in the case of the Greek Island of Chios, strategic planning must involve all stakeholders to ensure sustainability and balanced development. Action plans can facilitate the strategic planning process as the activities are broken down into time-bound blocks of actions. The development of action plans must include local stakeholders (Mason, 2020). Action plans must consider items that are within the control of the implementation agency.

10.4 Spatial and physical planning

The type of tourism planning activities engaged by a tourism department may not involve the details of spatial and physical planning. Spatial and physical planning are the processes of arranging the layout of development plans. It is an intuitive and a scientific process that involves tried and tested layout techniques. Nevertheless, this type of planning is essential for sustainable development as a goal and critical for island destinations, since based on the size of an island, space constraints limit tourism activities. Dede and Ayten (2012) have constructed a framework that includes national strategies and plans, environmental order plans, development and construction plans, urban design, and sustainable architectural design. Spatial and physical plans must be made implementable to realize sustainable development. One of the implementation issues of this planning approach is the legal issues that arise when there are conflicts between plans in the same area that may result in unplanned development (Dede & Ayten, 2012). An integrated approach to planning and institutional enforcements must be coordinated. The need for spatial planning enforcements on the Canary Islands, for example Law 6 (2001), stopped construction of hotels with lower star ratings (Báez-García, Flores-Muñoz, & Gutiérrez-Barroso, 2018). Enforcement of a legal framework is necessary to achieve sustainable development. Understanding responses to enforcement are also important. As such, social research is just as important to the spatial and physical planning processes as engineering and architectural activities. As a formal process, spatial planning endeavours to develop along a grid using methods such as zoning and land use.

Zoning involves assigning an area to be used for a specific purpose. The zones are determined by several factors such as geology, ecology, human pressures, morphology, and topography. Xie, Chandra, and Gu (2013) have identified morphological changes in coastal tourism on Denarau Island, Fiji, and have noted the importance of spatial planning in the identification of suitable areas for tourism development and protected zones. Zoning regulations specify land use, and such regulations are critical for the preservation of areas. Hernández-Delgado et al. (2012) have noted increasing construction in ecologically sensitive areas, and this practice is non-sustainable. Land use planning frameworks must be integrative and inclusive with greater involvement of local communities (Hernández-Delgado et al., 2012). Land use for tourism purposes is often not specifically defined, however, commercial use can cater for the development of a hotel or built attraction. Island tourism development is concentrated in coastal areas (Cave & Brown, 2012), and therefore the challenges of climate change impacts are more profound. Resources that may be used for tourism activities also include vast expanses of natural land areas and marine resources. Herein is the complication of spatial and physical planning and the interaction with sustainable development.

Natural attributes are particularly important for island destination products and services. Moon and Han (2018) have suggested that certain island destination attributes such as the natural environment and convenience of travelling on island contribute to tourist satisfaction and result in a revisit intention. Yang, Ge, Ge, Xi, and Li (2016) have found that tourism development is predicated on the basis of changes in land use, and tourist resources and tourist transportation are supporting factors for growth and development. Island tourism development faces two main sustainable development issues, as suggested by Kerr (2005): (1) scale of natural and human resources, and (2) isolation making the cost of transport and manufacturing expensive. Issues about fresh water, waste generation, environmental controls, and land use are particularly important for the sustainable development of islands. Nicely and Palakurthi (2012) have pointed out the stark negative environmental impacts on islands from tourism activities such as the limited freshwater supply and the improper waste disposal that contaminates the water supply.

Once a natural resource is destroyed, it cannot be regained as it will be lost or degraded. Kelman (2019, p. 408) has warned about the 'eco-island trap', as real sustainability is presented and not achieved. Spatial and physical plans must consider how resources will be used and the growth in use of resources to achieve sustainability. The term 'overtourism' comes up when overuse directly impacts sustainability of resources. Rather than viewing 'overtourism' as a negative term, it should be viewed in the context as suggested by Scheyvens and Momsen (2008) as control 'over tourism' to bring about positive effects in local communities. A likely outcome of overuse is decline of the very resource being used. Sustainable development means there is a balance between use and preservation, and such a balance

---

Best Practice 1: The islands of Tahiti sustainable tourism roadmap

An example of a sustainable tourism plan is the five-year Fāri'ira'a Manihini 2027 of Tahiti, French Polynesia, with capping the number of tourists to no more than the island inhabitants at any point in time (Hutcheon, 2022). The plan promises to be a sustainable tourism planning model. In 2022, the islands welcomed 242,907 visitors, an increase of 194% when compared with 82,546 visitors in 2021, and in 2022 receipts from tourism was US$ 0.71 billion and average spend per visitor was US$ 3,253.33 (not including international flights) (Ministry of Tourism French Polynesia, 2023). The islands of Tahiti plan to preserve the islands' cultural heritage and sustainability while allowing visitors to have an authentic experience of Polynesian culture (Tahiti Tourisme, 2022). The ministry of tourism, French Polynesia, has set out the new plan based on the experiences of the last global pandemic in 2020. The plan notes stimulation and adaptation goals while anticipating, responding, and relating to the needs of the islands (Ministry of Tourism French Polynesia, 2022). At the heart of the roadmap is "the strengthening of the authentic and sincere bond between those who come from overseas and those here, who consent to welcome them" (Ministry of Tourism French Polynesia, 2022, p. 3). The four major strategies are thoughtful development, strengthening the social bond, affirming cultural identity, and ensuring effective and collaborative government (Ministry of Tourism French Polynesia, 2022).

---

must be in the planning framework for spatial and physical planning activities. For example, beaches are a major attractor to islands, and as a result a large part of tourism infrastructure has developed along the coastline. Coastal development requires careful spatial and physical planning to cater for the physical forces of nature including: water, wind, and waves.

10.5 Project planning and outcome mapping methodology

In terms of planning frameworks, project plans are detailed action plans with a budget assigned. A distinction must be made between a project and a programme. A project has an end date, whereas a programme could be ongoing with no end date. A project must include the inputs of all stakeholders who will benefit from the project. Beyer, Anda, Elber, Revell, and Spring (2005) have advised that project plan development must balance indigenous interests, including a collaborative engagement strategy, risk management, and training and development opportunities. A project monitoring and evaluation plan assists with achieving project objectives. Two of the challenges of implementing projects are keeping the project within timeframe and within budget. Project management tools and techniques, for example, Gantt charts, cost–benefit analysis, feasibility studies, and PERT (programme evaluation and review technique), assist with keeping a project on target. Contingencies are built in the project plan to cater for any exigencies. One of the drawbacks of using a project planning approach for sustainable tourism development is that it is standalone, and may not be integrated with other projects within the same area. Nonetheless, a best practice in community-based tourism is allowing the community to express feedback about the project plan (Junaid, Sigala, & Banchit, 2021). Projects within the same community should be carefully coordinated, particularly if these are government projects to avoid wastage. Outcome mapping methodology (OMM) as a project planning approach improves stakeholder mapping by the identification of boundary and strategic partners who are integral to project implementation success.

OMM is a project or programme planning framework that achieves development results (Earl, Carden, & Smutylo, 2001). OMM involves various stages and steps that monitor the progress of the project at every stage (Figure 10.1). The process starts with visioning the outcome or impact of the project. OMM achieves strategic purpose through the involvement of boundary and strategic partners. OMM focuses on behavioural changes of boundary partners who are the ones that will benefit from the project. Strategic partners are also identified to support project activities. Once project markers are set, the success of the project is monitored

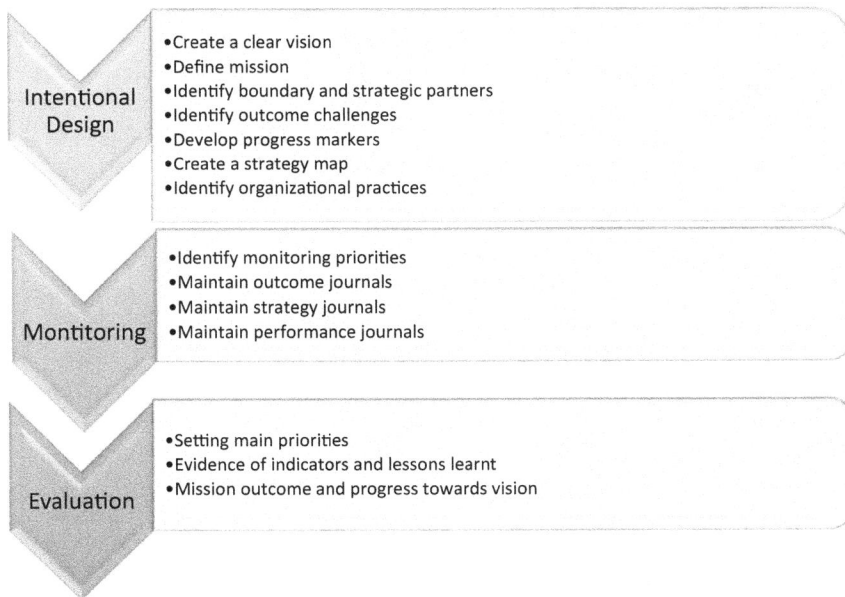

Figure 10.1 Stages in outcome mapping methodology.

Source: Modified from Earl, Carden, and Smutylo (2001).

using three journals: outcome journal, strategy journal, and performance journal (Earl et al., 2001). As boundary partners build capacity, results are realized through progress markers of 'expect to see', 'like to see', and 'love to see' (Earl et al., 2001). Changing organizational practices to suit the needs of the project and obtaining ongoing feedback help with project implementation. OMM tests theories of change to realize development impacts that are evaluated at the end of the project (McNaughton, McLeod, McNaughton, & Walcott, 2016). Outcome mapping is being popularized in measuring the benefits of tourism in islands. Warne and Thompson (2022) have developed a context, mechanism, and outcome approach for evaluating the benefits of tourism interventions in the Solomon Islands.

10.6 Conclusion

Planning for sustainable island tourism development can take various directions. The most appropriate approach depends on the availability of resources for planning process activities. Primarily tourism development requires the involvement of stakeholders in planning activities (Özgit & Öztüren, 2021). Such an approach garners support for plan implementation. In addition, the tourism planning process can be hindered by a lack of government commitment and support (Gurtner, 2016). Tourism planning can also become a political process as government officials seek to address endemic problems and issues in island destinations. Bearing this in mind, it is proposed that the most effective type of planning approach must be adopted to achieve results. Based on the exploration of tourism planning, sustainable development in island destinations can benefit from longer-term planning periods with scenario planning and outcome mapping elements. Planning is not impossible, but a workable or implementable plan must be carved out from a range of competing priorities and interests. As a result, ongoing progress and dynamic changes in the tourism environment must be integrated in planning activities.

Chapter 10 discussion questions

1 Select any island sustainable tourism plan and identify the data and information needed to develop this plan.
2 Discuss the advantages and disadvantages of three planning frameworks in the context of sustainable development.
3 Apply one planning framework to develop a sustainable development plan for an island destination.

References

Alberts, A., & Baldacchino, G. (2017). Resilience and tourism in islands: Insights from the Caribbean. In R. Butler (Ed.), *Tourism and resilience* (pp. 150–162). Wallingford, UK: CABI.

Azzopardi, E., & Nash, R. (2016). A framework for island destination competitiveness – Perspectives from the island of Malta. *Current Issues in Tourism, 19*(3), 253–281.

Báez-García, A. J., Flores-Muñoz, F., & Gutiérrez-Barroso, J. (2018). Maturity in competing tourism destinations: The case of Tenerife. *Tourism Review, 73*(3), 359–373.

Beyer, D., Anda, M., Elber, B., Revell, G., & Spring, F. (2005). *Best practice model for low-impact nature-based sustainable tourism facilities in remote areas.* Queensland: Sustainable Tourism Cooperative Research Centre.

Canavan, B. (2014). Sustainable tourism: Development, decline and de-growth. Management issues from the Isle of Man. *Journal of Sustainable Tourism, 22*(1), 127–147.

Caribbean Tourism Organization (2020). *Caribbean sustainable tourism policy and development framework.* Retrieved from Barbados: https://ourtourism.onecaribbean.org/resour ces/caribbean-sustainable-tourism-policy-framework-2020/

Cave, J., & Brown, K. G. (2012). Island tourism: Destinations: An editorial introduction to the special issue. *International Journal of Culture, Tourism and Hospitality Research, 6*(2), 95–113.

Croes, R. (2022). *Small island and small destination tourism: Overcoming the smallness barrier for economic growth and tourism competitiveness.* Abingdon, UK: CRC Press.

Dede, O. M., & Ayten, A. M. (2012). The role of spatial planning for sustainable tourism development: A theoretical model for Turkey. *Tourism: An International Interdisciplinary Journal, 60*(4), 431–445.

Earl, S., Carden, F., & Smutylo, T. (2001). *Outcome mapping: Building learning and reflection into development programs.* Ottawa, ON, Canada: International Development Research Centre.

Firdaus, N., & Endah, N. H. (2015). Accelerating the development of Bangka Island through sustainable tourism by strengthening the roles of multi-stakeholder. *International Journal of Administrative Science & Organization, 22*(3), 169–179.

Gurtner, Y. (2016). Returning to paradise: Investigating issues of tourism crisis and disaster recovery on the island of Bali. *Journal of Hospitality and Tourism Management, 28*, 11–19.

Hernández-Delgado, E. A., Ramos-Scharrón, C. E., Guerrero-Pérez, C. R., Lucking, M. A., Laureano, R., Méndez-Lázaro, P. A., & Meléndez-Díaz, J. O. (2012). Long-term impacts of non-sustainable tourism and urban development in small tropical islands coastal habitats in a changing climate: Lessons learned from Puerto Rico. *Visions for Global Tourism Industry-Creating and Sustaining Competitive Strategies, InTech Publications, Rijeka*, 357–398.

Hutcheon, H. (2022). The Islands of Tahiti details five-year sustainable tourism plan. *Seatrade Cruise News.* Retrieved from www.seatrade-cruise.com/ports-destinations/isla nds-tahiti-details-five-year-sustainable-tourism-plan

Junaid, I., Sigala, M., & Banchit, A. (2021). Implementing community-based tourism (CBT): Lessons learnt and implications by involving students in a CBT project in Laelae Island, Indonesia. *Journal of Hospitality, Leisure, Sport & Tourism Education, 29*, 100295.

Kelman, I. (2019). Critiques of island sustainability in tourism. *Tourism Geographies, 23*(3), 397–414.

Kerr, S. A. (2005). What is small island sustainable development about? *Ocean & Coastal Management, 48*(7–8), 503–524.

King, B., McVey, M., & Simmons, D. (2000). A societal marketing approach to national tourism planning: Evidence from the South Pacific. *Tourism Management, 21*(4), 407–416.

Lopes, V., Moreno Pires, S., & Costa, R. (2020). A strategy for a sustainable tourism development of the Greek Island of Chios. *Tourism: An International Interdisciplinary Journal, 68*(3), 243–260.

Mai, T., & Smith, C. (2018). Scenario-based planning for tourism development using system dynamic modelling: A case study of Cat Ba Island, Vietnam. *Tourism Management, 68*(October), 336–354.

Mason, P. (2020). *Tourism impacts, planning and management.* Oxon: Routledge.

McLennan, C.-l., Pham, T. D., Ruhanen, L., Ritchie, B. W., & Moyle, B. (2012). Counterfactual scenario planning for long-range sustainable local-level tourism transformation. *Journal of Sustainable Tourism, 20*(6), 801–822.

McLeod, M. (2023). Managing Caribbean tourism data ecosystems. *Current Issues in Tourism,* 1–17. doi:10.1080/13683500.2023.2288665

McLeod, M., Dodds, R., & Butler, R. (2021). Introduction to special issue on island tourism resilience. *Tourism Geographies, 23*(3), 361–370.

McNaughton, M., McLeod, M., McNaughton, M., & Walcott, J. (2016). *Open data as a catalyst for problem solving: Empirical evidence from a small island developing states (SIDS) context.* Paper presented at the Proceedings of the Open Data Research Symposium.

Ministry of Tourism French Polynesia (2022). *Fāri'ira'a Manihini 2027: The welcome that reflects us and binds us together, tourism development strategy for French Polynesia 2022–2027.* Tahiti: Ministry of Tourism French Polynesia. Retrieved from www.calameo.com/read/00346150392c0b997b7c5

Ministry of Tourism French Polynesia (2023). Key statistics and data. Retrieved from https://tahititourisme.org/en-org/market-sectors-and-statistics/key-statistics-and-data/

Moon, H., & Han, H. (2018). Destination attributes influencing Chinese travelers' perceptions of experience quality and intentions for island tourism: A case of Jeju Island. *Tourism Management Perspectives, 28,* 71–82.

Movono, A., & Hughes, E. (2022). Tourism partnerships: Localizing the SDG agenda in Fiji. *Journal of Sustainable Tourism, 30*(10), 2318–2332.

Nicely, A., & Palakurthi, R. (2012). Navigating through tourism options: An island perspective. *International Journal of Culture, Tourism and Hospitality Research, 6*(2), 133–144.

Özgit, H., & Öztüren, A. (2021). Conclusion: How could tourism planners and policymakers overcome the barriers to sustainable tourism development in the small island developing state of North Cyprus? *Worldwide Hospitality and Tourism Themes, 13*(4), 545–552.

Page, S. J., Yeoman, I., Connell, J., & Greenwood, C. (2010). Scenario planning as a tool to understand uncertainty in tourism: The example of transport and tourism in Scotland in 2025. *Current Issues in Tourism, 13*(2), 99–137.

Pizzitutti, F., Walsh, S. J., Rindfuss, R. R., Gunter, R., Quiroga, D., Tippett, R., & Mena, C. F. (2017). Scenario planning for tourism management: A participatory and system dynamics model applied to the Galapagos Islands of Ecuador. *Journal of Sustainable Tourism, 25*(8), 1117–1137.

Ponte, J. C., Couto, G., Pimentel, P., Sousa, Á., & Oliveira, A. (2021). Municipal tourism planning in an island territory: The case of Ribeira Grande in the Azores. *Tourism Planning & Development, 18*(3), 340–364.

Pyke, J., Lindsay-Smith, G., Gamage, A., Shaikh, S., Nguyen, V. K., de Lacy, T., & Porter, C. (2021). Building destination resilience to multiple crises to secure tourism's future. *Asia Pacific Journal of Tourism Research, 26*(11), 1225–1243.

Roxas, F. M. Y., Rivera, J. P. R., & Gutierrez, E. L. M. (2020). Framework for creating sustainable tourism using systems thinking. *Current Issues in Tourism, 23*(3), 280–296.

Scheyvens, R., & Momsen, J. (2008). Tourism in small island states: From vulnerability to strengths. *Journal of Sustainable Tourism, 16*(5), 491–510.

Seyitoğlu, F., & Costa, C. (2022). A systematic review of scenario planning studies in tourism and hospitality research. *Journal of Policy Research in Tourism, Leisure and Events*, 1–18.

Siti-Nabiha, A., & Saad, N. (2015). Tourism planning and stakeholders' engagements: The case of Penang Island. *Problems and Perspectives in Management, 13*(2), 269–276.

Spencer, A. J., Lewis-Cameron, A., Roberts, S., Walker, T. B., Watson, B., & McBean, L. M. (2023). Post-independence challenges for Caribbean tourism development: A solution-driven approach through Agenda 2030. *Tourism Review, 78*(2), 580–613.

Tahiti Tourisme (2022). The islands of Tahiti publishes a strategic roadmap for inclusive and sustainable tourism [Press release]. Retrieved from https://tahititourisme.org/en-org/press-releases/media-and-press/the-islands-of-tahiti-publishes-a-strategic-roadmap-for-inclusive-and-sustainable-tourism/

Teniwut, W. A., Hamid, S. K., & Makailipessy, M. M. (2022). Developing a masterplan for a sustainable marine sector in a small islands region: Integrated MCE spatial analysis for decision making. *Land Use Policy, 122*, 106356.

Warne, S. J., & Thompson, M. (2022). Future approaches to evaluating tourism in the developing world: Assessing realism in the Solomon Islands. *Journal of Hospitality and Tourism Management, 50*, 391–399.

Xie, P. F., Chandra, V., & Gu, K. (2013). Morphological changes of coastal tourism: A case study of Denarau Island, Fiji. *Tourism Management Perspectives, 5*, 75–83.

Yang, J., Ge, Y., Ge, Q., Xi, J., & Li, X. (2016). Determinants of island tourism development: The example of Dachangshan Island. *Tourism Management, 55*, 261–271.

11 Tourism sustainability resources

11.1 Introduction

The sustainability of island tourism destinations requires elaboration. An island is an ideal setting to explore and understand issues around sustainability (Burbano, Valdivieso, Izurieta, Meredith, & Ferri, 2022; McLeod, Dodds, & Butler, 2021). The tourism sector has ongoing challenges to constantly adapt to changes in the global environment. The identification and monitoring of the progress of sustainability indicators are promising activities to achieve sustainable development of island nations. More work is needed to identify and categorize sustainability indicators in an island context. While several studies have been conducted about sustainable tourism, the research has been largely a single case study based on the application of concepts to local and regional areas. A multiple case study approach can provide a comparative basis for understanding sustainable tourism in an island context.

A systems perspective of tourism activities aids in the understanding of the intricacies of sustainability as the inputs, processes, and outputs are broken down. Mai and Smith (2015) have utilized systems thinking to identify underlying system structures that may undermine sustainability. Issues around freshwater, land use, infrastructure, and human resources were identified in the case of Cat Ba Island (Mai & Smith, 2015). An approach is needed to address those issues affecting the sustainability of the tourism industry in islands. Predominantly the research surrounding sustainable tourism has focused on the sustainability pillars: economic, environmental, and social (Polnyotee & Thadaniti, 2014). Research is needed about the contexts of tourism development, particularly island countries that are threatened by overdevelopment and carrying capacity issues. Issues around the provision of dwellings for local communities in highly developed resort areas require attention (Cole & Razak, 2011). Small island destinations such as Aruba are faced with finding a balance between tourism development and the benefits gained by the local communities that support tourism development (Cole & Razak, 2011; Ridderstaat, Croes, & Nijkamp, 2016).

DOI: 10.4324/9781003435112-14

11.2 Tourism sustainability indicators

Islands require indicators for sustainable tourism industries (Cheer, 2020). Dymond (1997) has noted differences in the application of sustainability indicators. Font et al. (2021) have suggested that expectations for sustainability indicator implementation were unrealistic. Uses of sustainable tourism indicators include formulation of action plans, identification of short-term strategies, and benchmarking purposes (Lozano-Oyola, Blancas, González, & Caballero, 2012). Torres-Delgado and Saarinen (2014) have suggested that the effectiveness of sustainability indicators depends on the context. In that regard, certain conditions of a good sustainability indicator were outlined including adaptability, communication, comparability, cost, measurability, participation, precision, relevance, representation, sensitivity, and updating (Torres-Delgado & Saarinen, 2014). In support of the implementation of the indicators in the tourism sector, Font et al. (2021) have applied an absorptive capacity concept through acquisition and assimilation aspects to support the transfer of sustainable tourism indicator knowledge to the policymakers in the tourism sector. Sustainable tourism is a panacea of sorts with its links to sustainable development and destination resilience. A resilience concept has been applied to understand the sustainability of tourism. Destination resilience may be built through environmental sustainability indicators. Environmental sustainability indicators include energy management, expenditure on and protection of the natural ecosystems, the intensity of tourist use, pollution, water management, wastewater and solid waste management, and visual impacts on the built environment (Blancas, Lozano-Oyola, González, & Caballero, 2016).

Sustainable tourism indicators are the keys to unlocking sustainable tourism development that would benefit the country. A distinction has been made between indicators and measures, wherein a measure determines the extent of an occurrence. With sustainability indicators, the scope of activity may result in varying effects and longer-term impacts. First, measures in themselves become complicated as extensive lists have been found in the literature (Alfaro Navarro, Andrés Martínez, & Mondéjar Jiménez, 2020; Blancas, González, Lozano-Oyola, & Pérez, 2010; Miller & Ward, 2005). Second, measures should be categorized within a framework of indicators and not as a standalone consideration of sustainability. Third, measures have limitations in terms of time constraints, changes, variabilities, and interdependencies. For example, seasonality is a key concept in tourism, one measure in the peak season will have a different result in the off-peak season. Fourth, measures are context specific as these may vary between developed and developing countries, rural and urban tourist destinations, island and non-island tourist destinations. Given the challenges of measures and the broad requirement for the development of a sustainability framework, this chapter focuses on indicators and resources of sustainable tourism in islands.

Substantial work has been conducted on the indicators of sustainable tourism (Agyeiwaah, McKercher, & Suntikul, 2017; Blancas et al., 2016; Burghelea,

Uzlău, & Ene, 2016; Font et al., 2021; Ivars-Baidal, Vera-Rebollo, Perles-Ribes, Femenia-Serra, & Celdrán-Bernabeu, 2021; Lee & Hsieh, 2016; Rasoolimanesh, Ramakrishna, Hall, Esfandiar, & Seyfi, 2023). Early authors focused on the pillars of sustainability (Blancas et al., 2010). Lee and Hsieh (2016) have found 141 indicators for sustainable wetland tourism and have recommended the need for greater stakeholder involvement and collaboration in planning for sustainable tourism development. Important indicators on the list included respect for local culture, lifestyle, compliance, traffic problems, environment destruction, crowd management, local culture appreciation, the cost–benefit ratio, and reduction of environmental impact (Lee & Hsieh, 2016). Economic sustainability indicators include destination competitiveness, development control, economic benefits, seasonality, tourism employment, tourist experiences, tourist satisfaction, and transport (Blancas et al., 2016).

Social sustainability indicators are particularly important as these affect the well-being of the local community. Deery, Jago, and Fredline (2005) have considered the development of social indicators to determine tourism's impact on local communities and have noted the challenge of determining social indicator changes over time. Social indicators have been separated from cultural indicators in some instances (Agyeiwaah et al., 2017). Blancas et al. (2016) have identified several aspects of the social dimension of tourism sustainability including social carrying capacity and sociocultural effects on the host community. Sustainable tourism indicators are needed to monitor the performance as an appropriate means to achieve sustainability and resiliency of the tourism sector. In that regard, the selection and implementation of indicators should follow a cycle with individual indicators being part of a wider group of indicators (Blackstock, McCrum, Scott, & White, 2006). Indicators are then monitored based on an alert of 'Pressure' such as an increase in visitation, obtaining the 'State' by collecting data, determining the 'Impact' such as congestion or erosion, and finally implementing a 'Response' such as visitor management (Blackstock et al., 2006). While this approach is likely to yield the desired results of sustainable tourism, the volume and varied numbers of sustainable tourism indicators make such an approach burdensome. An assessment of the resources to realize sustainable tourism and resilience in island destinations must be done. For example, the sustainability of a tourist market after a natural disaster is of prime concern as there will be a need to rebuild physical infrastructure to restart tourism activities (McLeod, 2022).

11.3 Tourism sustainability aspects

Lee and Hsieh (2016) and Mai and Smith (2015) have identified sustainability indicators in the Cigu wetland, Taiwan, and Cat Ba Island, Vietnam, respectively, using a case study approach. One advantage of a case study approach is method flexibility in determining the best methods to achieve the research aim and objectives. For this chapter, the sustainability indicators and resources were identified from several cases in the literature. Secondary sources can be used to collect data about sustainable tourism indicators (Agyeiwaah et al., 2017; Deery et al., 2005). The data collection started with the selection of the sustainability indicators source documents. A total of 76 papers were selected using certain keywords about sustainable tourism in island destinations. These keyword phrases included sustainable tourism metrics, tourism stakeholder involvement, tourism metrics, factors influencing and barriers to sustainable tourism, implementing sustainable tourism, failures in sustainable tourism, and sustainable tourism indicators. The first step was a one-off search of the documents based on the relevance of the title and by browsing the abstract and overall content. The second step was to search for all content with indicators of sustainable tourism.

The final step was to select those papers that contained sustainability indicators and resources about island tourism. Documents were then reviewed to derive the sustainability indicators and these indicators were entered in a Microsoft Excel spreadsheet. Papers about sustainability indicators on a site level as a hotel were not included, for example, in an article by Muangasame and McKercher (2015). The selected indicators for the sustainability of the island broadly included several works (Altinay, Var, Hines, & Hussain, 2007; Cantallops, 2004; Cole & Razak, 2011; Farsari & Prastacos, 2000; Kunasekaran et al., 2017; Polnyotee & Thadaniti, 2014; Reddy, 2008; Tsaur & Wang, 2007; Twining-Ward & Butler, 2002; UNWTO, 2004). A total of 322 sustainability indicators and resources were identified from these papers with a maximum of 52 indicators and a minimum of 9 indicators per source. The dates of the papers ranged from 2000 to 2022.

Figure 11.1 illustrates the location of the 12 study settings across the globe and Table 11.1 the characteristics of the case studies. Table 11.1 shows 195 sustainability indicators and resources in settings with small islands and 107 sustainability indicators and resources in settings with large islands. Each island context is different in terms of geographical location (Figure 11.1) and political and socio-cultural circumstances (Table 11.1). In some cases, the islands are part of a larger jurisdiction or standalone, and these contexts have implications for the sustainable development of the tourism sector.

A key method was the development of a coding system for the sustainability indicators. To identify the nodes in the network, a three-tier coding was done (Table 11.2). Codes with five letters were developed. First, the first two letters of the code represented the sustainability pillars: economic (EC), environmental (EN), and socio-cultural (SO). Second, it was determined that the indicator may contribute positively (P) or negatively (N) to sustainability goals as adopted by several authors (Blancas et al., 2016; Lozano-Oyola et al., 2012). Third, the last two letters represented a specific indicator identification. Coding of the last two letters was time-consuming as the coding was done manually, and similar characteristics were grouped and re-grouped. For example, government agencies and various tourism businesses were regrouped into 'GA' and 'TB' codes, respectively. Indicator screening and fine-tuning are appropriate methods to identify and categorize sustainability indicators and have been previously done by Roberts and Tribe (2008). A total of 33 sustainability indicator groups were identified and aligned with the Sustainable Development Goals (SDGs).

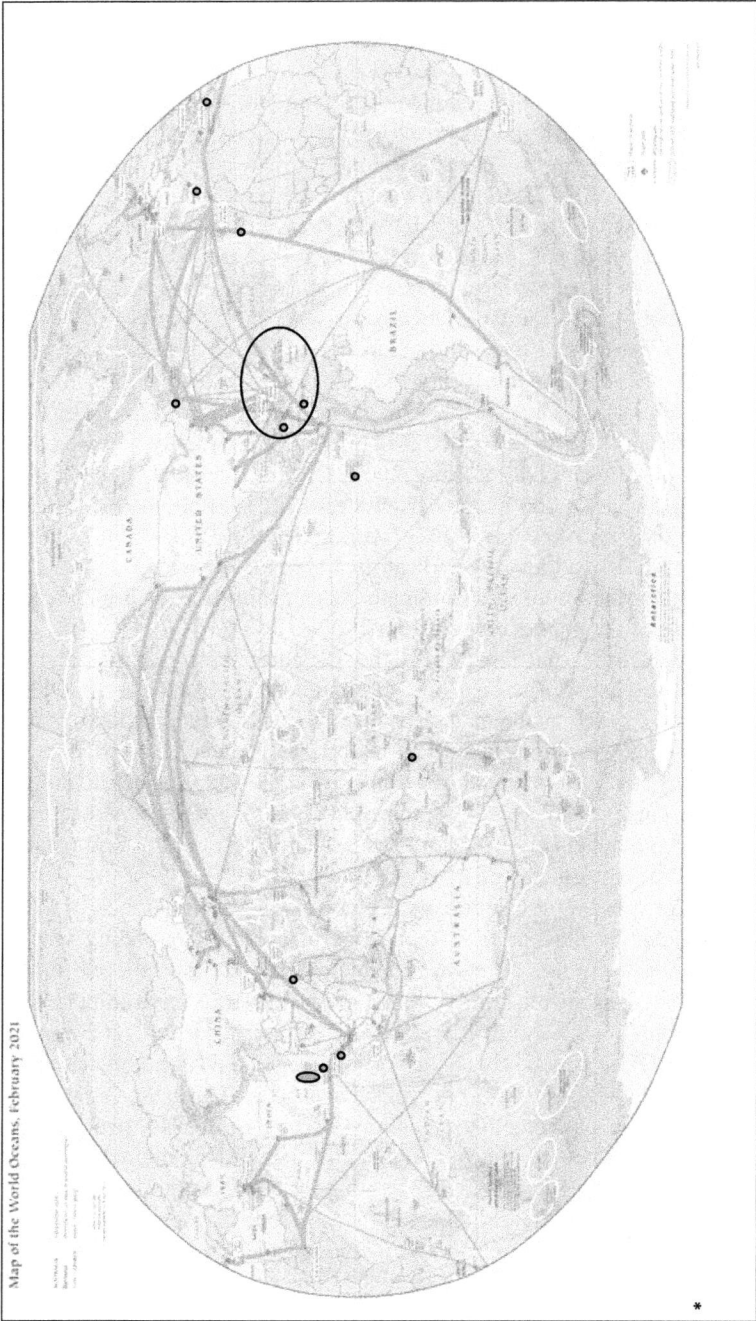

Figure 11.1 World oceans with research study islands and areas circled.

Source: CIA World Fact Book (Central Intelligence Agency, 2023).

Table 11.1 Island Case Studies Characteristics

Country/ region	Island/archipelago	Island/territory governance	Island area	Island population		Indicators
India	Andaman and Nicobar Islands	Union territory	8,249 km² (3,185 mi²)	379,944	2011	7
Netherlands	Aruba	Independent territory	180 km² (mi²)	115,600	2021	10
Spain	Balearic Islands	Parliamentary assembly	4,992 km² (1,927 mi²)	1,030,650	2007	**55**
Spain	Canary Islands	Independent territory	7,447 km² (2,875 mi²)	2,153,389	2019	**9**
Canada	Cape Breton Island	Municipality	10,311 km² (3,981 mi²)	132,019	2021	30
Caribbean	Caribbean Islands	Mixed governance	2,754,000 km² (1,063,000 mi²)	38,000,000	2017	**14**
Ecuador	Galapagos Islands	Provincial government	8,010 km² (3,093 mi²)	25,124	2010	44
Taiwan	Green Island	Township	15 km² (6 mi²)	3,000	Estimation	28
Crete	Hersonissos	Local government	272 km² (104.9 mi²)	21,135	2011	52
Jamaica	Jamaica	Independent country	10,830 km² (mi²)	2,827,695	2021	**29**
Thailand	Phuket Island	Local government	543 km² (210 mi²)	79,755	2018	24
Samoa	Samoa	Independent country	2,830 km² (mi²)	218,764	2021	20

Sources: Cape Breton Regional Municipality (2023), CIVITAS Initiative (2023), Ministry of the Interior (2023), Taiwanese Secrets (2023), ThoughtCo (2023), and World Bank (2023).

Note: Large islands, territories, and archipelagos are bolded indicator numbers.

Table 11.2 Island Sustainability Indicator Codes

Sustainability pillars		Sustainability direction		Sustainability indicators			SDGs
				Code	Resources	Additional content	
EC	Economic	N	Negative	BT	Beach per tourist	Maximum number of visitors per day; available beach surface; blue flag	15LL
EN	Environmental	P	Positive	CD	Cultural dilution	Cultural resources; cultural conservation; promotion of culture; special interest cultural sites	11SC
SO	Socio-cultural			EA	Environmental awareness	Environmental research, ecological assets	13CA
				EG	Environment energy	Electricity consumption; energy consumption per tourist; ecological footstep of energy consumption; energy intensity index; energy consumption; energy consumption and management; supply of energy; renewal energies index	7AE
				EP	Environmental projects	Environmental sustainability; environmental management systems; management and conservation; participation in land conservation; control of environmental impacts; control area access; monitoring and site control; residents knowledgeable about use controls; zoning regulation; evolution of natural protected area; cost to repair damaged sites; community-based projects; maximum concentration of numbers on fragile sites; impacts from tourism; plastic management	15LL
				ER	Environment resources	Endangered species; quantity of species; renewable resources; ecological and agricultural; replanted and reforested surfaces; vegetation cover; biodiversity; CO_2; air pollution; land degradation; spoiled coastline; forest fires; noise pollution	15LL; 13CA

EW	Waste generation	Solid waste treated; solid waste in the dump; garbage accumulation; wastewater treated; wastewater; waste production; wastewater treatment systems; solid waste treatment systems; waste treatment capacity	6CS; 12RC
FI	Funding	Financial resources; interest rates; financial transparency; efficient use of economic resources	8DW
FX	Taxes	Hotel tax; Ecotax	8DW
GA	Government agencies	Government support; government corruption; dependency on government	17PG
HQ	Hotel quality	Hotel footprint; occupancy; all-inclusive resorts; hotel capacity; new rooms; green; dwelling area; bars per population; hotels treating waste; winter/summer beds; beds with conference facilities; area conference capacity	9II
IR	Infrastructure resources	Roads; parking; concrete; rest stops; toilet; public transportation; traffic system; congestion; signs; number of vehicles in use; public health system; urbanization	11SC
IV	Investor	Investment; international resources; public–private partnership	8DW
LB	Labour	Labour balance; females/males; winter/summer population; population growth; unemployment rate and working population; high unemployment; area unemployment rate; human resources pressure index; informal employment	5GE, 8DW
LG	Legislation	Green Act; tourism law	16PI
LU	Land use	Open space to build for tourism; tourism and support land use area; dwelling area, associated land; territory and land use; tourist use/total use; dwelling area/hotel room; workers dwelling; tourist landscape under threat from development	15LL
LS	Living standard	Housing access and prices, household income; cost of living & conditions	11SC

(*Continued*)

Table 11.2 (Continued)

Sustainability pillars	Sustainability direction	Sustainability indicators		Additional content	SDGs
		Code	Resources		
		MG	Effective management	Use controls; administration; common goals	11SC
		PL	Planning	All islands; site plans; area covered; monitoring	11SC
		RM	Recycling material	Recycled water; waste recycling; water recycling; recycling and reduction	12RC
		SC	Stakeholder collaboration	Community involvement; local meeting to discuss; participation; tourism isolation; trust in stakeholders; less corruption and more transparency among stakeholders; addressing community concerns	17PG
		SF (CR)	Safety	Crime; security; disaster alarm; drugs; personal safety; accidents; visitor harassment; CCTV cameras; frequency of criminality; child prostitution	16PI, 3GW
		SI	Social impact	Improve the quality of life; socio-cultural sustainability; projects involving society participation	3GW
		TA	Tourist attraction	Recreation facilities; festivals; handicraft; tourist information	12RC
		TB	Tourism business	Local services; local products; local ownership; sustainable practices; ownership of tourism firms; illegal fishing; price control	1AP, 2EH, 8DW
		TE	Tourism employment	Employees graduate of tourism schools; unemployment rate off-season periods; job stability; labour seasonality; number of employed females/males; number of employed guides/interpreters; self-employment/family business; locals employed in tourism; tourism staff; foreign workers' problems; ecotourism jobs outside peak season; wages evaluation; wages evolution; salary	1AP, 2EH, 8DW

TM	Tourism model	Tourism pressure; importance of tourism; tourist efficiency index; tourist saturation; balance between high-end and informal tourism; carrying capacity	12RC
TO	Tourism operators	Tour operator; boats; moorings; tour guide	8DW, 12RC, 14LW
TP	Tourism passengers	Tourist arrivals; sea arrivals; site visits; like to return; fewer tourists; visitors to a specific site; visitors to sensitive sites; visitor behaviour; tourist participation in damaging activities; tourist perception; ecotourism experiences; local complaints about intrusions; tourist satisfaction; visitor preference; access for locals to visitor sites	8DW, 12RC, 14LW
TR	Tourism revenue	Foreign exchange; economic sustainability; earnings from tourists; linkages; tourism integration in the local economy; foreign trade; wealth distribution; resident satisfaction	8DW
TS	Tourism studies	Technical expertise; tourism research; impact studies; environmental studies; perception surveys	9II
TT	Tourism training	Capacity building; hotel staff going on training courses; short-term training courses; number of training courses; self-reliance; tourism awareness; stakeholder communication; English; education	4QE
WT	Water	Water quality; water analysis; water consumption per tourist; access to drinking water; water ration; water monitoring; water swimming; water swimming limits	6CS

A mechanism whereby sustainable development can be monitored should not only relate to the type and direction of sustainability indicators. Unlike other approaches used to adopt sustainability indicators in tourism, the adoption of the SDGs has the additional advantage of integration with national development. The SDGs that focus on decent work and economic growth (8DW) and quality education (4QE) are particularly important to strengthen the tourism sector. Industry innovation (9II) through the creation of tourism business opportunities will also address some of the socio-cultural matters around the well-being of host communities. Environmental indicators are particularly important to conserve the resources needed to support the tourism sector. SDG number 6, clean water and sanitation, is perhaps the most critical SDG to ensure the sustainability of the tourism sector as several indicators about waste generation, recycling, and water were revealed (Table 11.2). The role of the partnership SDG, 17PG, should not be underestimated as several issues about trust and transparency were noted in the indicator content (Table 11.2). Stakeholder collaboration is a must for successful tourism (McComb, Boyd, & Boluk, 2017; Nyanjom, Boxall, & Slaven, 2018).

11.4 Conclusion

Policies to facilitate sustainable tourism development in islands must consider the specific resources that are the foundation of the tourism industry. Burbano and Meredith (2021) have noted the need for targeted policy and regulations to contribute to sustainable outcomes in the Galápagos Islands. One of the challenges of achieving sustainable development is an understanding of the interrelationships of the economic, socio-cultural, and environmental aspects that are comprised on an island. These aspects vary from island to island, however, a framework that supports sustainability may be found by maximizing the benefits of positive aspects and minimizing the costs of negative aspects of island tourism. Exploring the case of Mauritius, Fauzel and Tandrayen-Ragoobur (2023) have pointed out the need for continual monitoring and instituting preventive measures to achieve sustainable tourism.

Islands' vulnerabilities must be placed in a context. Climate change, economic shifts, and human resource challenges will continue to evolve (Jones, 2014). Based on these contexts, sustainable development seems far-fetched for islands that are limited to find the supporting resources (Harrison & Pratt, 2015). External influences can easily subsume island government plans for development. Conditionalities mean that island governments' actions are constrained to achieve sustainable development (Bethell-Bennett, 2022). Based on such circumstances some mapping of internal resources must be conducted to understand the regenerative and inclusive aspects of island sustainability and resiliency. One way that sustainable development can be mapped and understood is using indicators. Indicators are elements that place the island on the right path or track towards achieving goals for sustainable development.

Chapter 11 discussion questions

1 Explain the environmental, economic, and socio-cultural resources that are needed to support sustainable tourism.
2 Discuss the view that island destinations are sustainable based on the number of positive sustainability indicators.
3 Apply the island sustainability codes to determine the existing resources in any island destination.

References

Agyeiwaah, E., McKercher, B., & Suntikul, W. (2017). Identifying core indicators of sustainable tourism: A path forward? *Tourism Management Perspectives, 24*(October), 26–33.

Alfaro Navarro, J.-L., Andrés Martínez, M.-E., & Mondéjar Jiménez, J.-A. (2020). An approach to measuring sustainable tourism at the local level in Europe. *Current Issues in Tourism, 23*(4), 423–437.

Altinay, L., Var, T., Hines, S., & Hussain, K. (2007). Barriers to sustainable tourism development in Jamaica. *Tourism Analysis, 12*(1–2), 1–13.

Bethell-Bennett, I. (2022). When storms strike: Performing tourism, hurricanes, and a pandemic in accumulation and dispossession. In I. Bethell-Bennett, S. Rolle, J. Minnis, & F. Okumus (Eds.), *Pandemics, disasters, sustainability, tourism* (pp. 193–210). Leeds, UK: Emerald Publishing.

Blackstock, K., McCrum, G., Scott, A., & White, V. (2006). *A framework for developing indicators of sustainable tourism.* Retrieved from: https://macaulay.webarchive.hutton. ac.uk/ruralsustainability/FrameworkReport.pdf

Blancas, F. J., González, M., Lozano-Oyola, M., & Pérez, F. (2010). The assessment of sustainable tourism: Application to Spanish coastal destinations. *Ecological Indicators, 10*(2), 484–492.

Blancas, F. J., Lozano-Oyola, M., González, M., & Caballero, R. (2016). Sustainable tourism composite indicators: A dynamic evaluation to manage changes in sustainability. *Journal of Sustainable Tourism, 24*(10), 1403–1424.

Burbano, D. V., & Meredith, T. C. (2021). Effects of tourism growth in a UNESCO World Heritage Site: Resource-based livelihood diversification in the Galapagos Islands, Ecuador. *Journal of Sustainable Tourism, 29*(8), 1270–1289.

Burbano, D. V., Valdivieso, J. C., Izurieta, J. C., Meredith, T. C., & Ferri, D. Q. (2022). "Rethink and reset" tourism in the Galapagos Islands: Stakeholders' views on the sustainability of tourism development. *Annals of Tourism Research Empirical Insights, 3*(2), 100057. https://doi.org/10.1016/j.annale.2022.100057

Burghelea, C., Uzlău, C., & Ene, C. M. (2016). Comparative indicators of sustainable tourism. *Scientific Papers Series Management, Economic Engineering in Agriculture and Rural Development, 16*(3), 77.

Cantallops, A. S. (2004). Policies supporting sustainable tourism development in the Balearic Islands: The Ecotax. *Anatolia, 15*(1), 39–56.

Cape Breton Regional Municipality (2023). Cape Breton Island. Retrieved from www.cbrm. ns.ca/

Central Intelligence Agency (2023). The world factbook. Retrieved from www.cia.gov/the-world-factbook/static/b8c037fed6ac6c04eba873188effbc39/world_oceans.pdf

Cheer, M. (2020). The urgency for sustainability indicators. *Institute of Island Studies.* Retrieved from https://islandstudies.com/files/2022/08/Annual-Report-on-Global-Isla nds-2019-Chapter-5-Tourism-on-small-islands-The-urgency-for-sustainability-indicat ors-Joseph-M.-Cheer.pdf

CIVITAS Initiative (2023). Hersonissos (Greece). Retrieved from https://civitas.eu/cities/ hersonissos#

Cole, S., & Razak, V. (2011). Island awash–sustainability indicators and social complexity in the Caribbean. In M. Budruk & R. Phillips (Eds.), *Quality-of-life community indicators for parks, recreation and tourism management* (pp. 141–161). London: Springer.

Deery, M., Jago, L., & Fredline, L. (2005). A framework for the development of social and socioeconomic indicators for sustainable tourism in communities. *Tourism Review International, 9*(1), 69–77.

Dymond, S. J. (1997). Indicators of sustainable tourism in New Zealand: A local government perspective. *Journal of Sustainable Tourism, 5*(4), 279–293.

Encyclopædia Britannica (2023). Geography & travel. Retrieved from www.britannica.com/

Farsari, Y., & Prastacos, P. (2000). *Sustainable tourism indicators: Pilot estimation for the municipality of Hersonissos, Crete. International Scientific Conference "Tourism on Islands and Specific Destinations"* Chios 14–16 December 2000. Retrieved from Sustainable Tourism Indicators: Pilot Estimation for the Municipality of Hersonissos (diva-portal.org).

Fauzel, S., & Tandrayen-Ragoobur, V. (2023). Sustainable development and tourism growth in an island economy: A dynamic investigation. *Journal of Policy Research in Tourism, Leisure and Events, 15*(4), 502–512.

Font, X., Torres-Delgado, A., Crabolu, G., Palomo Martinez, J., Kantenbacher, J., & Miller, G. (2021). The impact of sustainable tourism indicators on destination competitiveness: The European tourism indicator system. *Journal of Sustainable Tourism, 31*(7), 1608–1630.

Harrison, D., & Pratt, S. (2015). Tourism in Pacific island countries: Current issues and future challenges. In S. Pratt & D. Harrison (Eds.), *Tourism in Pacific Islands: Current issues and future challenges* (pp. 3–21). Abingdon: Routledge.

Ivars-Baidal, J. A., Vera-Rebollo, J. F., Perles-Ribes, J., Femenia-Serra, F., & Celdrán-Bernabeu, M. A. (2021). Sustainable tourism indicators: What's new within the smart city/destination approach? *Journal of Sustainable Tourism, 31*(7), 1556–1582.

Jones, A. (2014). Global environmental change and small island states and territories: Economic and labour market implications of climate change on the tourism sector of the Maltese Islands. Retrieved from www.um.edu.mt/library/oar/handle/123456 789/17193

Kunasekaran, P., Gill, S. S., Ramachandran, S., Shuib, A., Baum, T., & Herman Mohammad Afandi, S. (2017). Measuring sustainable indigenous tourism indicators: A case of Mah Meri ethnic group in Carey Island, Malaysia. *Sustainability, 9*(7), 1256.

Lee, T. H., & Hsieh, H.-P. (2016). Indicators of sustainable tourism: A case study from a Taiwan's wetland. *Ecological Indicators, 67*(August), 779–787.

Lozano-Oyola, M., Blancas, F. J., González, M., & Caballero, R. (2012). Sustainable tourism indicators as planning tools in cultural destinations. *Ecological Indicators, 18*(July), 659–675.

Mai, T., & Smith, C. (2015). Addressing the threats to tourism sustainability using systems thinking: A case study of Cat Ba Island, Vietnam. *Journal of Sustainable Tourism, 23*(10), 1504–1528.

McComb, E. J., Boyd, S., & Boluk, K. (2017). Stakeholder collaboration: A means to the success of rural tourism destinations? A critical evaluation of the existence of stakeholder collaboration within the Mournes, Northern Ireland. *Tourism and Hospitality Research, 17*(3), 286–297.

McLeod, M. (2022). Tourism destination recovery, a case study of Grand Bahama Island. In I. Bethell-Bennett, S. Rolle, J. Minnis, & F. Okumus (Eds.), *Pandemics, disasters, sustainability, tourism* (pp. 93–108). Leeds, UK: Emerald Publishing.

McLeod, M., Dodds, R., & Butler, R. (2021). Introduction to special issue on island tourism resilience. *Tourism Geographies, 23*(3), 361–370.

Miller, G., & Ward, L. T. (2005). *Monitoring for a sustainable tourism transition. The challenge of developing & using indicators.* Wallingford, UK: CABI.

Ministry of the Interior (2023). Green Island: The vitality and crisis of the Tiny Heaven. Retrieved from https://np.cpami.gov.tw/np-quarterly.html?view=endetail&catid=17&id=139&ls=0

Muangasame, K., & McKercher, B. (2015). The challenge of implementing sustainable tourism policy: A 360-degree assessment of Thailand's "7 Greens sustainable tourism policy". *Journal of Sustainable Tourism, 23*(4), 497–516.

Nyanjom, J., Boxall, K., & Slaven, J. (2018). Towards inclusive tourism? Stakeholder collaboration in the development of accessible tourism. *Tourism Geographies, 20*(4), 675–697.

Polnyotee, M., & Thadaniti, S. (2014). The survey of factors influencing sustainable tourism at Patong beach, Phuket Island, Thailand. *Mediterranean Journal of Social Sciences, 5*(9), 650–650.

Rasoolimanesh, S. M., Ramakrishna, S., Hall, C. M., Esfandiar, K., & Seyfi, S. (2023). A systematic scoping review of sustainable tourism indicators in relation to the sustainable development goals. *Journal of Sustainable Tourism, 31*(7), 1497–1517. doi:10.1080/09669582.2020.1775621

Reddy, M. V. (2008). Sustainable tourism rapid indicators for less-developed islands: An economic perspective. *International Journal of Tourism Research, 10*(6), 557–576.

Ridderstaat, J., Croes, R., & Nijkamp, P. (2016). The tourism development–quality of life nexus in a small island destination. *Journal of Travel Research, 55*(1), 79–94.

Roberts, S., & Tribe, J. (2008). Sustainability indicators for small tourism enterprises – An exploratory perspective. *Journal of Sustainable Tourism, 16*(5), 575–594.

Taiwanese Secrets (2023). Green Island Taiwan travel guide. Retrieved from www.taiwanese-secrets.com/green-island-taiwan/

ThoughtCo (2023). Caribbean countries by land area. Retrieved from www.thoughtco.com/caribbean-countries-by-area-4169407

Torres-Delgado, A., & Saarinen, J. (2014). Using indicators to assess sustainable tourism development: A review. *Tourism Geographies, 16*(1), 31–47.

Tsaur, S.-H., & Wang, C.-H. (2007). The evaluation of sustainable tourism development by analytic hierarchy process and fuzzy set theory: An empirical study on the Green Island in Taiwan. *Asia Pacific Journal of Tourism Research, 12*(2), 127–145.

Twining-Ward, L., & Butler, R. (2002). Implementing STD on a small island: Development and use of sustainable tourism development indicators in Samoa. *Journal of Sustainable Tourism, 10*(5), 363–387.

UNWTO (2004). *Indicators of sustainable development for tourism destinations.* Madrid: World Tourism Organization.

World Bank (2023). World development indicators data bank. Retrieved from https://databank.worldbank.org/source/world-development-indicators

12 Sustainable island tourism development

12.1 Introduction

The sustainability of tourism is based on interconnected actions that will facilitate the ongoing functioning of a tourism system. On the demand side, tourism sustainability is the ongoing flow of tourists to a destination area. On the supply side, tourism sustainability is the ongoing provision of tourism products and services. The tourism sector is dynamic, and tourism activities are complicated with links between a tourist-generating market through a transit route to a tourist destination area (Leiper, 1979). As such, Burghelea, Uzlău, and Ene (2016) have pointed out the complexity of the tourism system that makes the identification of sustainable development strategies difficult. A variety of methods has been applied to understand sustainable development but a holistic method is still to be found. Epistemological challenges about understanding sustainable tourism development in islands can be overcomed with the application of social network analysis (SNA). Social network analysis was used to map island sustainability indicators. This methodology allowed for the development of a system of codes and then the connections were made using the island case studies. Network relationships among items show actors, which are nodes, and the connections, which are ties, forming a network structure. It is this structure that is measured to understand the position of an actor and the value of that actor to all other actors in the network structure. Network measurement to understand the bonding of the nodes in the network structure may also be conducted.

Sustainable tourism development in islands can be achieved through a focus strategy on the key resources of sustainability. Network structures were constructed using UCINET 6 software (Borgatti, Everett, & Freeman, 2002) and illustrated using NetDraw software (Borgatti, 2002). An island sustainability indicator network is a whole network that is static, showing the relationship of the indicators within and among the islands. Two types of networks of sustainability indicators were constructed. The first type of network shows the islands or territories and the sustainability indicators that were identified for each island. The second type of network was the sustainability indicators only. For the network with the indicators only it was particularly important to identify a focal node for

DOI: 10.4324/9781003435112-15

each island case study since the island name node was removed as the focal node. In the first instance, the degree centrality values were calculated for the sustainability indicators to determine the more common nodes of island sustainability. However, rather than use the centrality values as the basis for selecting the focal nodes, a random number was used to select the focal node from each island case study node list. This is because the indicators with higher centrality values that became focal nodes skewed the network structure. By constructing the network of sustainability indicators using randomly selected focal nodes, the network pattern may be viewed as a social construction. Such a method supports the view that sustainability may be socially constructed, as the importance of particular sustainability indicators depends on the society in which those indicators have been identified (Cole & Razak, 2011).

Once the network structure was constructed, for the first type of network, whole network density measures were calculated. Density is the ratio of actual ties to all possible ties (McLeod, 2014) and indicates the convergence of the various sustainability indicators. The second type of network, in addition to the overall network, was divided into larger and smaller islands, and hypothesis tests were conducted to determine network heterogeneity and homogeneity. Measurement of the island sustainability indicators network characteristics and hypotheses indicates certain outcomes of sustainable tourism development. For example, network heterogeneity means that positive indicators are interacting with negative indicators and coordination is required for network outcomes. Network theory has implications for the sustainable development of islands. Convergence has been identified as a network coordination characteristic that bonds similar nodes or actors together (Borgatti & Halgin, 2011). Convergence indicates an extent of homogeneity in the network structure and in this case that the islands are taking similar sustainability paths. Borgatti and Halgin (2011) have argued that the homogeneity of structural environments results in the sameness of outcomes. Quadratic assignment procedure correlation tests can be done to test hypotheses about the degree of homogeneity and heterogeneity in the network structure (McLeod, 2023). An elaboration of the systems thinking of sustainable tourism development in islands have been expounded in the following sections.

12.2 Tourism sustainability systems

One approach to understanding how a sustainable tourism system functions is to develop indicators. Indicators are pointers to some aspect of the system that requires monitoring to achieve certain outcomes. Herein is the value of network theory. A theory of networks posits that the structure and dynamics of a network determine outcomes (Borgatti & Halgin, 2011). Based on this premise, the mapping of nodes or actors, and the ties or relationships, allows for determining that structure. The underlying reasons for the ties or flows in the network involve hypothesis testing to reveal the attributes that determine the connections. This chapter applies network theory to a network of island sustainability indicators.

The indicators are those that were identified in island case studies. As the indicators were coded, those indicators with the same code are interconnected forming a network component of island sustainability indicators. An examination of 322 sustainability indicators and the interconnectedness of these indicators are the antecedents of island tourism sustainable development. The first five letters of the islands were included as the code for each island. For example, Jamaica was coded as JAMAI.

Understanding the network structure involves several viewpoints. First, the position of the nodes' centrality may influence network outcomes (Ledesma Gonzalez, Merinero-Rodríguez, & Pulido-Fernández, 2021). Second, as there are no missing ties since the indicators were determined using case studies, whole network measures are used to determine the structural characteristics of the network. Third, the explanation of the flows, ties, or patterns relates to the underlying attributes of the nodes (McLeod, 2020b). Two attributes were tested: an indicator type attribute (economic, environmental, and socio-cultural) and an indicator direction attribute (positive and negative). If the direction or type explains the ties, then this results in certain outcomes of the network structure and dynamics. The direction or type of indicator as an antecedent of network consequences of island sustainability requires elaboration.

Network theory posits that network homogeneity means that the same attributes determine the network pattern (Borgatti & Halgin, 2011). A statistically significant homogeneous network with indicator direction will result in sustainable (positive direction) or unsustainable (negative direction) island tourism. Similarly, a statistically significant heterogeneous network with positive or negative network direction will result in unsustainable island tourism. In contrast, network heterogeneity of island sustainability indicators, based on indicator type is good, as economic, environmental, and socio-cultural indicators build island tourism resilience and sustainability. In island sustainability theory, an island destination with sustainability indicators that are statistically significantly homogeneous in positive indicators and heterogeneous in indicator type will be more sustainable and resilient. The implications relate to building network homogeneity and heterogeneity of island tourism sustainability indicators that will result in sustainable development.

12.3 Sustainable tourism development in islands

The island sustainability indicators were interconnected in a network structure. The island tourism sustainability network consists of nodes (islands and indicators) and ties (the relationship between the islands and indicators). The relational data were developed by using an island as a focal node in the network and then the indicators related to the focal node as ties. The indicators are also interconnected by an indicator being connected to the same indicator in other islands. For the network structure with islands, indicator nodes and ties, the indicator characteristics were described based on the indicator's network position, and network structural characteristics were determined by whole network density measures.

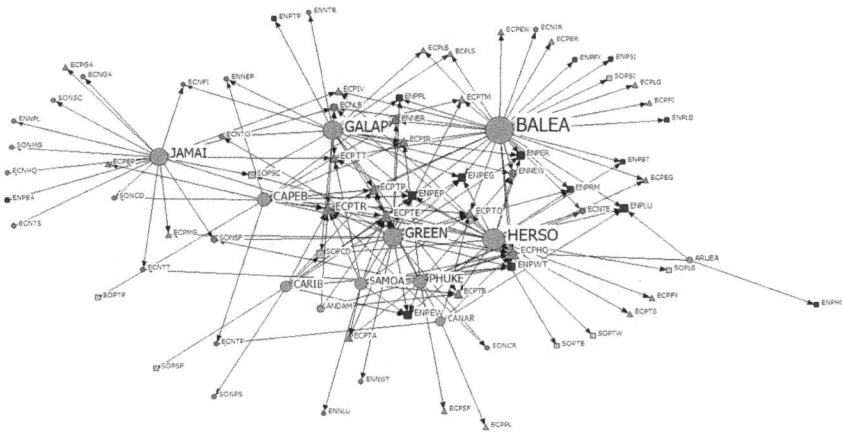

Figure 12.1 Island sustainability indicators network (degree centrality).

Notes: Islands, circles sized by degree centrality; economic, light up triangle; environment, dark rounded square; socio-cultural, square; all negative indicators, circle.

The whole network measures revealed a network of 83 nodes and 181 ties with an average degree of 2.181 and a density of 0.027 (Figure 12.1). An average degree of 2 means that each indicator node is related to two other indicator or island nodes on average. Given a network of 83 nodes, an overall density of 0.027 indicates that 2.7% of all possible ties exist. A degree centrality measure was used to represent the size of the island nodes. Island tourism sustainability and resiliency are built on the sustainability indicator types and positive or negative directions. Five central indicators are evident. In terms of environmental sustainability indicators, these are ENPEP – environment positive environmental projects, and ENPEG – environmental positive energy. In terms of economic sustainability the indicators are ECPTE – economic positive tourism employment, ECPTP – economic positive tourist passengers, ECPTR – economic positive tourism revenue (Figure 12.1) (Burbano, Valdivieso, Izurieta, Meredith, & Ferri, 2022; Farsari & Prastacos, 2000; Polnyotee & Thadaniti, 2014; Tsaur & Wang, 2007; UNWTO, 2004). The ECPTP – economic positive tourism passengers sustainability indicator is very important for island destinations (UNWTO, 2004). Visitation to sensitive sites, visitor behaviour, tourist participation in damaging activities, tourist perception, ecotourism experiences, local complaints about intrusions, and tourist satisfaction are important aspects of tourism demand that must be understood and managed. Environmental projects, energy, tourism employment, and tourism revenue are the other key sustainability indicators for tourism in islands (Cantallops, 2004; Farsari & Prastacos, 2000; Tsaur & Wang, 2007).

Island tourism features are important for understanding island sustainability and resilience. Islands positioned to the right such as the Balearics Islands and Hersonissos have more positive indicators and the Galapagos Islands have been positioned in the middle and closer to Jamaica because of the presence of some negative sustainability indicators. The Jamaica case study was about barriers to tourism sustainability (Altinay, Var, Hines, & Hussain, 2007). The proximity of five of the island cases and one area is also indicative of the similarity of sustainability indicators, Phuket Island, Andaman and Nicobar Islands, Samoa, Green Island, Canary, and the Caribbean. Sharpley (2012) has concurred that islands share similar characteristics regarding production and consumption of tourism activities but has concluded that island tourism practices are not distinctive from other forms of tourism. In the context of sustainable development, the positioning of these islands on a network map of sustainability indicators suggests that there are distinctive sustainability challenges that islands have that must be addressed. Island tourism lessons are to be learnt about the resource needs and indicator monitoring to ensure sustainable development is achieved.

In addition, as the islands are in various geographical areas, indicator similarities suggest that the physical environmental characteristics of an island alone may not account for sustainable outcomes, but other socio-cultural and similar economic conditions are important. Differences exist for the economic, environmental, and socio-cultural sustainability indicators. Economic sustainability indicators dominate the network. The socio-cultural indicators are largely on the periphery, except in the case of those indicators connected to Jamaica and the group of islands in the south of the network structure. The environmental resource sustainability indicators are spread across the network, however, are mostly to the east of the network map, and include indicators that are not central such as water, waste generation, land use, and recycling. Several authors have identified water as a sustainability indicator (Burbano et al., 2022; Twining-Ward & Butler, 2002). An ordering of the environmental sustainability indicators is indicative of the important network position that the indicators have regarding sustainable tourism in the islands. The proximity of the environmental sustainability indicators to the Balearic Islands and Hersonissos, Crete, is an indication of the supporting policies that are needed to ensure sustainability of island tourism.

A group of sustainability indicators positioned between Jamaica and the Balearic Islands is also observed. In terms of the economic indicators, ECNLB – economic negative labour, ECPIV – economic positive investor, ECPIR – economic positive infrastructure resources, and ECPTT – economic positive tourism training (Altinay et al., 2007; Polnyotee & Thadaniti, 2014; UNWTO, 2004) were positioned near the Galapagos Islands on the sustainability indicator network. Labour, investment, training, and infrastructure contribute to island tourism economies and should be effectively monitored and managed, particularly the negative aspects of labour. Negative aspects of labour include high unemployment in industry and unemployment overall in the islands (Altinay et al., 2007). A second network structure was constructed to reveal sustainability indicator types (economic, environmental, and socio-cultural) and direction (positive or negative).

12.4 Tourism sustainability indicators' network

The whole network measures for the sustainability indicator network revealed 70 nodes with 163 ties and a density of 3.4% (Figure 12.2). The average degree centrality was 2.329. Nodes that are central to the sustainability indicators network are primarily related to the tourism sector, including ECPTE – economic positive tourism employment, ECPTP – economic positive tourism planning, ENPER – economic positive environmental resources, and environment positive water (Figure 12.2) (Farsari & Prastacos, 2000; Polnyotee & Thadaniti, 2014; Tsaur & Wang, 2007; UNWTO, 2004). Hotel quality, tourism models, tourism operators, and tourism attractions are also important sustainability indicators for islands (Cantallops, 2004; Twining-Ward & Butler, 2002). A node on the far left, SONCD relates to cultural dilution. The cultural dilution node forms part of the island case study about barriers to sustainable tourism noted by Altinay et al. (2007).

A further analysis was done to determine the sustainability indicators between the small and larger islands. Seven islands were in the small island group and five islands were in the large island group (see Table 11.1). The focal sustainability indicators for the large islands were mainly environmental, relating to awareness, tax, and planning; and the economic indicators were living standards and tour operators (Figure 12.3). Different results may have been obtained if the focal nodes were other than those randomly selected, however, as the sustainability network is constructed based on a random sample of nodes, this reduces the likelihood of error in understanding a network of sustainability indicators. The calculated whole network measures revealed a density of 2.4% with 53 nodes and 66 ties, and the average degree was 1.245 for the large islands' sustainability indicator network (Figure 12.3). The focal sustainability indicators for the small islands were

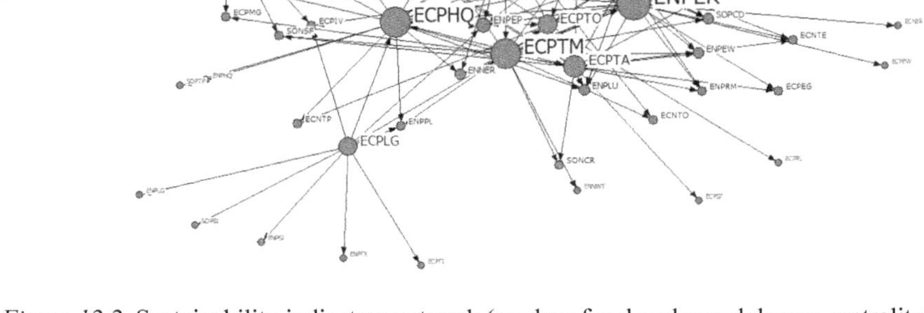

Figure 12.2 Sustainability indicators network (random focal nodes and degree centrality).

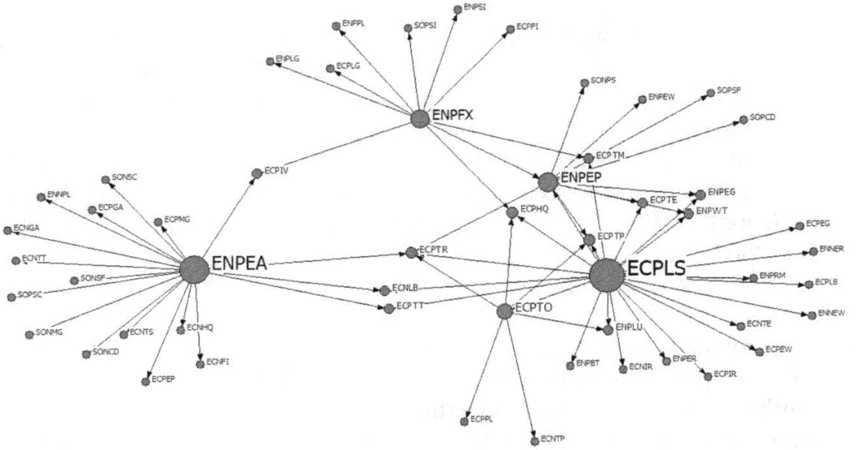

Figure 12.3 Sustainability indicators network for large islands (random focal nodes and degree centrality).

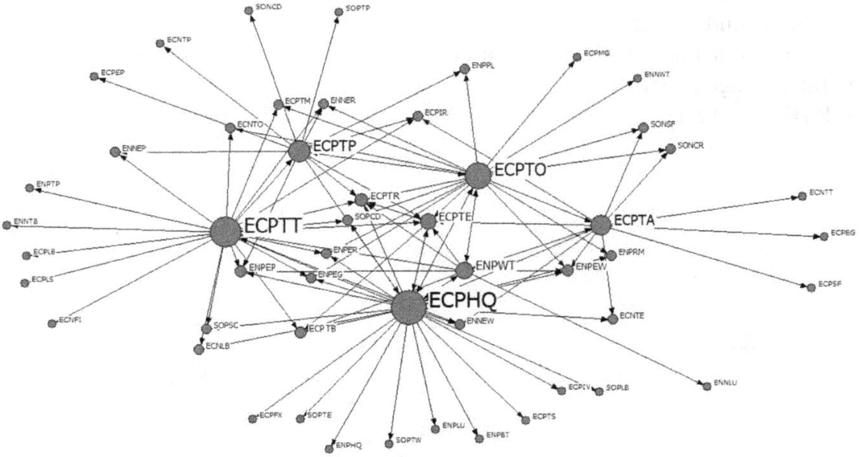

Figure 12.4 Sustainability indicators network for small islands (random focal nodes and degree centrality).

Table 12.1 Island Sustainability Indicators (Quadric Assignment Procedure Correlation Results)

Sustainability indicators		Heterogeneity (absolute difference)				Homogeneity (same)			
		ρ	p	Mean	SD	ρ	p	Mean	SD
Island sustainability indicators	Direction	-0.103	0.107	-0.001	0.081	0.103	0.102	-0.001	0.081
	Type	-0.000	0.501	-0.001	0.080	-0.032	0.342	0.001	0.080
Small island sustainability indicators	Direction	-0.009	0.473	-0.001	0.099	0.009	0.472	0.002	0.100
	Type	-0.173	**0.038***	-0.003	0.098	0.137	0.090	-0.001	0.099
Large island sustainability indicators	Direction	0.027	0.407	-0.003	0.132	-0.027	0.418	0.002	0.134
	Type	-0.084	0.275	0.000	0.135	0.047	0.363	-0.000	0.133

Notes: Calculated using UCINET 6 (Borgatti, Everett, & Freeman, 2002); ρ, Pearson correlation; p values bolded – *p < 0.05; SD, standard deviation.

mostly economic, relating to hotel quality, tourism training, tourism planning, tour operators, and tourism attractions (Figure 12.4). Tourism employment, tourism revenue, and water were centrally positioned in the sustainability indicator network for small islands (Figure 12.4). The whole network measures were calculated for the small islands' sustainability indicator network, and the results revealed a network containing 51 nodes with 101 ties, a density of 4%, and an average degree of 1.98 (Figure 12.4).

The results from the quadratic assignment procedure correlation analysis illustrate the extent of convergence or homogeneity in the sustainability indicators networks. While sustainability indicators having a positive or negative direction did not explain the network outcomes, the types of indicators whether economic, environmental, or socio-cultural did explain the network outcome for small islands. The heterogeneity results were statistically significant with $p = 0.038$ for indicator type for small islands (Table 12.1). This is an indication that having a balance between the types of sustainability indicators is important for small island tourism sustainability.

12.5 Tourism sustainability and resiliency framework

Several approaches to handling the sustainability of tourism were conducted. Blackstock, McCrum, Scott, and White (2006) have proposed the selection and implementation of indicators should follow an indicator cycle starting with a vision, choosing indicators, grouping indicators, applying and interpreting, and then re-visioning. Deery, Jago, and Fredline (2005) have noted an indicator development process to measure tourism impact. While these approaches have an advantage in the identification of indicators, the relationships between the indicators and the resultant outcomes of such relationships must be unpacked. The inputs of sustainability have been described, and the processes have been related to tourism governance structures. The concept of panarchy may assist with understanding sustainability as tourism may be viewed as an ecosystem of actors (McLeod, 2020a). To ensure that sustainable tourism indicators are identified and adaptable to various contexts, a comprehensive sustainable tourism indicator framework is needed.

The tourism sector is often a mainstay for island economies. Island sustainability is an important matter for all islands as these are more vulnerable to external shocks. This chapter has identified, categorized, and assessed island tourism sustainability indicators. The chapter advances understanding of the key issues around island tourism sustainability by comprehensively examining 322 indicators from 14 island case studies. Patterns have emerged that illustrate the more important sustainability indicators for island tourism. The idea that tourism bounces back, and returns to its normal state as in a recovery, means that it is sustainable. While the commonly used island sustainability indicators have been identified and categorized, the chapter also argues for the coordination of networks of sustainability indicators to realize sustainable development outcomes. Coordination of sustainability indicator types can result in sustainable development and resiliency in small islands.

Small islands have a greater network density of 4% and are statistically significantly heterogeneous in the capacity of indicators' types to achieve the sustainability

Figure 12.5 Small island tourism sustainability and resiliency framework.

of tourism. Mai and Smith (2015) have argued for the application of systems thinking to understand sustainable tourism in islands. This chapter clarifies such systems thinking by setting out the antecedents and consequences of sustainability indicators for sustainable tourism development. Small islands that focus on six sustainability indicators, hotel quality, tourism planning, water, tourism training, tour operators, and tourism attractions (Figure 12.5), and the subsequent network structure (Figure 12.4) can realize island sustainability and resiliency. Hotel quality, tour operators, and tourist attractions are important indicators of sustainability in islands (Cantallops, 2004; Farsari & Prastacos, 2000). The supply and consumption of water have been a concern in islands and affect the sustainable development of islands (Polnyotee & Thadaniti, 2014; UNWTO, 2004). Water recycling has been proposed as a solution to the problem of lack of water (UNWTO, 2004). Tourism planning on the macro and micro levels is necessary to support the sustainability of tourism activities on islands. In the Canary Islands, tourism development plans are needed (UNWTO, 2004), and environmental planning (Tsaur & Wang, 2007) and moving away from the traditional, centralized approach to planning (Altinay et al., 2007) have been noted regarding tourism planning. The underlying indicators that support the sustainability of tourism in small islands have been identified from a systems perspective.

This chapter illustrates the need for sustainability indicators. One group of indicators is not more important than the other. After careful selection of priority indicators, policy actions should be taken to address important indicators. To a greater extent, the sustainability indicators are predominantly economically related rather than environmentally and to a lesser extent socio-culturally related (Figure 12.1). Such a finding reflects the substantial economic importance of tourism activities on islands. In addition, the set of indicators is predominantly positive; however, negative indicators are to be effectively monitored, particularly those relating to the labour force. While positive indicators are to be encouraged, negative indicators should be reversed (Blancas, Lozano-Oyola, González, & Caballero, 2016; Lozano-Oyola, Blancas, González, & Caballero, 2012). Policies to address negative sustainability indicators augur well for an island destination. While the sustainable tourism literature addresses community tourism and community involvement, these results have not shown the prominence of communities as a sustainability indicator.

12.6 Conclusion

Interactions of island tourism sustainability indicators may determine certain outcomes for island sustainable development. Theoretically, heterogeneity of indicator types means that resilience building is supported by developing a balance of indicator types. Heterogeneity means that sustainability indicators of different types are interacting, and this brings about island tourism system balance. The workings of sustainable development must be unpacked to understand the likely outcomes from a range of policy actions towards sustainable development. Islands face a range of challenges that make understanding key priorities for sustainable development difficult. In addition, islands have different political, cultural, and social contexts, and therefore adjustments and modifications must be made to the sustainability indicators to achieve the Sustainable Development Goals.

In practicality, sustainability indicators appeared in several islands that were not geographically co-located. In addition, the network centrality results indicate those sustainability indicators that are more prominent in island tourism. Some islands were positioned closer to a group of environmental indicators and other islands were closer to economic indicators. This is an indication of the varied focus of islands when it comes to tourism sustainability indicators. This chapter provides a guide to the important sustainability indicators for sustainable development. Those islands that are threatened may utilize the small island sustainability indicator framework (Figure 12.5) and embark upon planning activities to improve island tourism sustainability and resiliency. The prominent indicators of island tourism sustainability are those economic and environmental indicators that were more central in Figure 12.1. Mapping of sustainability indicators assists with the prioritization of limited resources in islands, and as a result less important sustainability indicators may be defunded, and resources placed elsewhere to fund more significant island tourism sustainability indicators.

Chapter 12 discussion questions

1 Explain the concept of island tourism sustainability.
2 Discuss the viewpoint that sustainability indicators in large islands are different than those in small islands.
3 Apply the small island sustainability indicator framework to a small island tourism destination.

References

Altinay, L., Var, T., Hines, S., & Hussain, K. (2007). Barriers to sustainable tourism development in Jamaica. *Tourism Analysis, 12*(1–2), 1–13.

Blackstock, K., McCrum, G., Scott, A., & White, V. (2006). A framework for developing indicators of sustainable tourism. Retrieved from UK: https://macaulay.webarchive.hutton.ac.uk/ruralsustainability/FrameworkReport.pdf

Blancas, F. J., Lozano-Oyola, M., González, M., & Caballero, R. (2016). Sustainable tourism composite indicators: A dynamic evaluation to manage changes in sustainability. *Journal of Sustainable Tourism, 24*(10), 1403–1424.

Borgatti, S. P. (2002). *Netdraw network visualisation.* Harvard, MA: Analytic Technologies.

Borgatti, S. P., Everett, M. G., & Freeman, L. C. (2002). *Ucinet for Windows: Software for social network analysis.* Harvard, MA: Analytic Technologies.

Borgatti, S. P., & Halgin, D. S. (2011). On network theory. *Organization Science, 22*(5), 1168–1181.

Burbano, D. V., Valdivieso, J. C., Izurieta, J. C., Meredith, T. C., & Ferri, D. Q. (2022). "Rethink and reset" tourism in the Galapagos Islands: Stakeholders' views on the sustainability of tourism development. *Annals of Tourism Research Empirical Insights, 3*(2), 100057. https://doi.org/10.1016/j.annale.2022.100057

Burghelea, C., Uzlău, C., & Ene, C. M. (2016). Comparative indicators of sustainable tourism. *Scientific Papers Series Management, Economic Engineering in Agriculture and Rural Development, 16*(3), 77.

Cantallops, A. S. (2004). Policies supporting sustainable tourism development in the Balearic Islands: The Ecotax. *Anatolia, 15*(1), 39–56.

Cole, S., & Razak, V. (2011). Island awash–sustainability indicators and social complexity in the Caribbean. In M. Budruk & R. Phillips (Eds.), *Quality-of-life community indicators for parks, recreation and tourism management* (pp. 141–161). London: Springer.

Deery, M., Jago, L., & Fredline, L. (2005). A framework for the development of social and socioeconomic indicators for sustainable tourism in communities. *Tourism Review International, 9*(1), 69–77.

Farsari, Y., & Prastacos, P. (2000). Sustainable tourism indicators: Pilot estimation for the municipality of Hersonissos, Crete. International Scientific Conference "Tourism on Islands and Specific Destinations" Chios 14–16 December 2000. Retrieved from Sustainable Tourism Indicators: Pilot Estimation for the Municipality of Hersonissos (diva-portal.org).

Ledesma Gonzalez, O., Merinero-Rodríguez, R., & Pulido-Fernández, J. I. (2021). Tourist destination development and social network analysis: What does degree centrality contribute? *International Journal of Tourism Research, 23*(4), 652–666.

Leiper, N. (1979). The framework of tourism: Towards a definition of tourism, tourist, and the tourist industry. *Annals of Tourism Research, 6*(4), 390–407.

Lozano-Oyola, M., Blancas, F. J., González, M., & Caballero, R. (2012). Sustainable tourism indicators as planning tools in cultural destinations. *Ecological Indicators, 18*(July), 659–675.

Mai, T., & Smith, C. (2015). Addressing the threats to tourism sustainability using systems thinking: A case study of Cat Ba Island, Vietnam. *Journal of Sustainable Tourism, 23*(10), 1504–1528.

McLeod, M. (2014). Analysing inter-business knowledge sharing in the tourism sector. In M. McLeod & R. Vaughan *Knowledge networks and tourism* (pp. 143–157). Abingdon, UK: Routledge.

McLeod, M. (2020a). Tourism governance, panarchy and resilience in the Bahamas. In S. Rolle, J. Minnis, & I. Bethell-Bennett. *Tourism development, governance, and sustainability in the Bahamas* (pp. 103–113). Abingdon, UK: Routledge.

McLeod, M. (2020b). Understanding knowledge flows within a tourism destination network. *Journal of Hospitality and Tourism Insights, 3*(5), 549–566.

McLeod, M. (2023). Tourism policy networks in four Caribbean countries. *Annals of Tourism Research Empirical Insights, 4*(2), 100113.

Polnyotee, M., & Thadaniti, S. (2014). The survey of factors influencing sustainable tourism at Patong beach, Phuket Island, Thailand. *Mediterranean Journal of Social Sciences, 5*(9), 650–650.

Sharpley, R. (2012). Island tourism or tourism on islands? *Tourism Recreation Research, 37*(2), 167–172.

Tsaur, S.-H., & Wang, C.-H. (2007). The evaluation of sustainable tourism development by analytic hierarchy process and fuzzy set theory: An empirical study on the Green Island in Taiwan. *Asia Pacific Journal of Tourism Research, 12*(2), 127–145.

Twining-Ward, L., & Butler, R. (2002). Implementing STD on a small island: Development and use of sustainable tourism development indicators in Samoa. *Journal of Sustainable Tourism, 10*(5), 363–387.

UNWTO (2004). *Indicators of sustainable development for tourism destinations.* Madrid: World Tourism Organization.

13 Sustainable development futures

13.1 Introduction

Sustainable development suggests that development will be placed along a continuum that results in ongoing adaptation and sustainability (Lim & Cooper, 2009). The achievement of such an aim means that economic activities such as tourism will be supported and benefits derived. Island tourism as a form of economic activity requires special consideration. First, an island, particularly small islands, does not contain vast resources to support masses of tourists entering the island unless those resources can be readily imported. In addition, McElroy (2003) has pointed out that islands in the Pacific, Indian, and African oceans have been hindered in developing tourism based on their remoteness and lack of direct air connectivity. Second, outward oriented tourism has created a dependency that is unsustainable. Brohman (1996) has considered the challenges and problems with outward oriented tourism including foreign dependency, creation of enclaves, and spatial inequalities. Third, the extent of tourism activities in an island generates waste and requires energy use that an island cannot support sustainably. Kelman (2019) has suggested that islands import waste such as plastics and batteries, and waste left on beaches do not depict the clean surroundings necessary to support tourism activities. The future of island tourism rests in finding a balance between tourism activities and island sustainability.

The previous chapters in this third part of the book explored planning and methodology for understanding island tourism sustainability. This chapter sets out the main elements to facilitate a sustainable island tourism destination including the Sustainable Development Goals, sustainable green investment, sustainable system transformation, and sustainable and regenerative tourism. At the core of the process of development must be island sustainability for the local population. In that regard, a best practice case of Dominica about building a climate resilient island, has been included in this chapter. The stage of tourism development must be considered as tourism requires built infrastructure, and island governments may not have available financial resources or access to capital to meet the required development needs. Francis (2012), in a study about the Solomon Islands, has noted the need for financial support from local, national, and international agencies to provide the necessary green infrastructure. A wholistic sustainable development

DOI: 10.4324/9781003435112-16

plan may be a possible action, however, the need for spatial and physical plans that consider sustainable development matters, also requires resources that may not be indigenous to an island. One example is the Inter-American Development Bank's tourism strategy and action plan for Jamaica with strategic objectives to support rural development, innovation, and entrepreneurship (Inter-American Development Bank, 2018).

The future of sustainable development in islands must be mapped out based on the political, socio-cultural, as well as physical characteristics of an island. Connell (2018) has noted the conflicts among social, economic, and environmentally sustainable development objectives, and has argued for a balance between development and sustainability. The Sustainable Development Goals (SDGs) provide a guideline to achieve sustainable development, however, the goals are resource bound. In an island that is resource constrained, a framework is needed to ensure that sustainable development can be achieved. The interrelationships of the goals must be understood to minimize duplication of efforts in achieving multiple goals. Sachs et al. (2019) have simplified and have transformed the 17 SDGs into 6 transformations to achieve the goals. Support of the SDGs requires the necessary investment. Investment in renewable resources including the green economy and building climate resiliency must be evaluated in the context of an island. With the resources for the SDGs being prioritized and sourcing investments, system transformation and regenerative practices can be achieved.

13.2 Sustainable Development Goals (SDGs)

In the Caribbean, concerns about exceeding carrying capacity and overdevelopment in the tourism sector have been recorded (Cole & Razak, 2011). To determine a path that supports sustainability of island tourism destinations, literature about sustainable tourism indicators was reviewed and indicators compiled and categorized. Several authors have developed lists of sustainable tourism indicators and categories (Agyeiwaah, McKercher, & Suntikul, 2017; Blancas, Lozano-Oyola, González, & Caballero, 2016; Lee & Hsieh, 2016). As previously outlined in Table 11.2, the SDGs were matched with sustainability indicators. Social network analysis (SNA) (Borgatti, 2002; Borgatti, Everett, & Freeman, 2002) was conducted using the statistically significant small island data given in Table 12.1 to determine the relationships and the denser sustainability indicators and SDGs. Findings from the analysis suggest the sustainability indicators and SDGs that are dominant for island sustainability, and these findings have implications for prioritizing policy areas and resource allocations that support island tourism's sustainability and resiliency.

Analyses of the sustainability indicators and the SDGs provide substantial benefits to understanding tourism development. Indicators of sustainable tourism assist with simplifying the myriad of aspects that support or pose a barrier to tourism development (Altinay, Var, Hines, & Hussain, 2007; Ivars-Baidal, Vera-Rebollo, Perles-Ribes, Femenia-Serra, & Celdrán-Bernabeu, 2021). The SDGs are global initiatives for assessing national plans, and governments must report on the achievement of these goals. Several lessons can be learnt by adoption of the SDGs to the development of the tourism sector including managing tourism investment and business operations (Nair & McLeod, 2020). A framework that utilizes the SDGs will bring success for overall national development and facilitates understanding of the gaps towards the sustainability and resiliency of tourism. The challenge is finding a way to relate the SDGs to tourism activities in a manner that is accurate and relevant.

Further analysis was necessary to understand the SDGs in relation to the sustainability indicators. The sustainability indicators were interconnected and the economic indicators, tourism employment, tourism revenue, and tour operators, have a central position (Figure 13.1). The pattern does show the strength of the ties of the more important sustainability indicators. For example, ENPLU (environment positive land use node) was very important for the development of tourism in Aruba (Cole & Razak, 2011), however this indicator was linked to one other island, Hersonissos, Crete (Figure 13.1). Economic positive infrastructure (ECPIR) and economic positive tourism passengers (ECPTP) showed greater tie strength in the network than other sustainability indicators (Figure 13.1) (Polnyotee & Thadaniti, 2014; UNWTO, 2004). The environmental sustainability indicator with prominence was ENNEW – environment negative waste generation (Polnyotee & Thadaniti, 2014). A case studied by Tsaur and Wang (2007) in Green Island, Taiwan, is positioned central and closest to the positive economic indicators. While Hersonissos, Crete, and Galapagos Islands, Ecuador, have more indicators of sustainability, the position of the islands in the network map suggests different

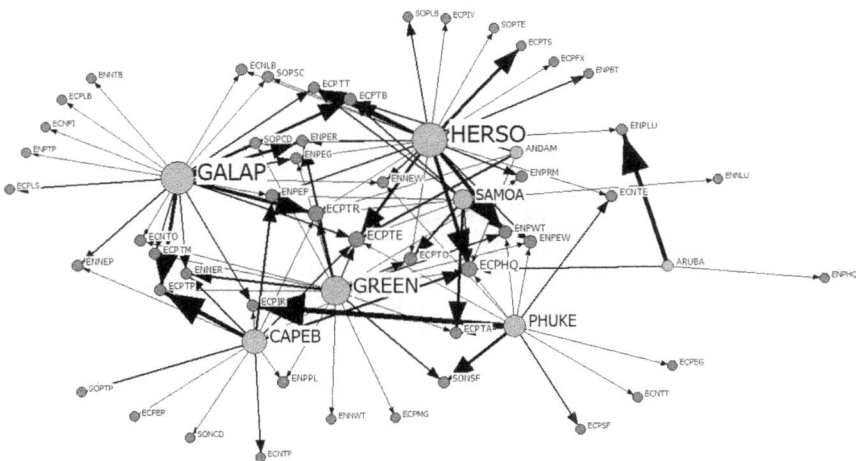

Figure 13.1 Small island tourism sustainability indicators network (degree centrality).

Notes: Data from Figure 12.1 and Table 11.2. (Islands – grey circle, indicators – grey circle, node size and labels equal degree centrality and line thickness equals tie strength.)

sustainability indicators. In the case of Hersonissos (Farsari & Prastacos, 2000), several socio-cultural indicators are evident, and in the case of the Galapagos Islands (Burbano, Valdivieso, Izurieta, Meredith, & Ferri, 2022), several negative environmental and economic indicators were evident.

The network of the SDGs was interconnected and it illustrated the more important SDGs for small island tourism sustainability including decent work (8DW), ending hunger (2EH), and alleviating poverty (1AP) (Figure 13.2). Responsible consumption (12RC) was a central goal in all cases except Aruba and Andaman and Nicobar Islands. While clean water and sanitation (6CS) and life on land were not central, these were important as sustainability indicators based on the tie strength (Figure 13.2). The SDG of industry and innovation (9II) was not central in the network, but had greatest tie strength to Hersonissos, Crete (Figure 13.2). In addition, the SDGs network indicates that SDG 10, reduce inequalities, was not represented by the sustainability indicators for small island tourism destinations.

Sustainable tourism can be achieved through the adoption of the SDGs. This multiple-case example illustrates a method for the development and achievement of sustainability and resiliency for island tourism destinations. A practical implication is the prioritization of SDGs that will enhance sustainable tourism and as a result the more important indicators are identified and managed. Achievement of the SDGs in island nations requires a consensus among government, private sector, and civil society. Each island has a different context that must be understood in relation to each goal. Identification of the goals and the progress made must be

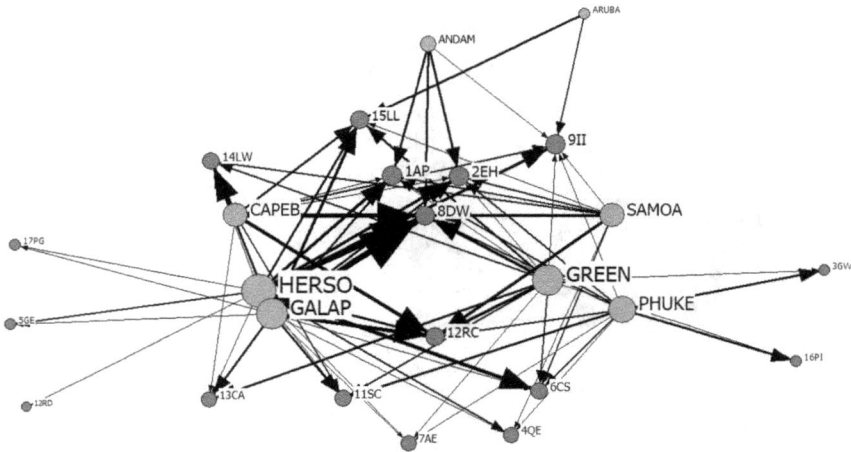

Figure 13.2 Small island tourism Sustainable Development Goals network (degree centrality).

Notes: Data from Figure 12.1 and Table 11.2. (Countries – grey circle, indicators – grey circle, node size and labels equal degree centrality and line thickness equals tie strength.)

monitored to achieve the SDGs and island tourism's sustainability and resiliency. With prioritization of the sustainability indicators and SDGs, policy formulation and implementation, as well as resources deployed, a sustainable development future in island nations is achieved.

13.3 Sustainable green investment

In the context of island tourism sustainable development, sustainable investment has received limited attention. This section explores ideas of green investment and climate financing to address sustainable development in island nations. Green investments are investment activities associated with protecting the environment (Siedschlag & Yan, 2021). A report of the Green Growth Action Alliance points out that green investment could be grown by focusing on developing country's needs (World Economic Forum, 2013). Green investment can be facilitated through macro-economic policies relating to interest and exchange rate facilities, cross-border lending facilities, and funds (United Nations Environment Programme, 2023).

Barbados has a climate finance initiative. Islands are impacted on both sides of the coin. Climate events have resulted in drought and desertification on some islands (McLeod, 2023). Natural disasters have threatened the economic resilience of islands, and these are related to climate-change consequences (Kennedy, Crawford, Main, Gauci, & Schembri, 2022). With slow economic growth and complicated recovery systems, climate finance is an urgent priority for island economies. Borrowing to recover is not a sustainable way forward. Linnerooth-Bayer, Mechler, and Hochrainer (2011) have pointed out the role of insurance against natural disasters in developing countries and the adoption of risk pooling and transfer strategies. Islands do not have the financial strength to handle climate-change-related impacts in most instances. The Bridgetown Initiative regarding climate financing calls for changing the terms and conditions of development banks (Agenda, 2023). Climate change is a global crisis, and policies and institutions must address this growing challenge to bring about a balance between countries that have the capabilities to deal with the climate crisis and countries that do not have the capability to handle these dynamic events.

The success of green investment must be considered within the context of islands with resource limitations. A green investment policy framework was developed and outlines infrastructure investment in energy, transport, and water (Corfee-Morlot et al., 2012). The framework incentivizes the private sector to invest in green infrastructure as low-carbon climate resilient infrastructure is constrained by market failure, investment barriers, and high upfront costs and requires long time periods (Corfee-Morlot et al., 2012). To hedge off and protect such investments, new approaches to investing in green infrastructure are needed. McKinsey and Company noted that private-sector-led investment in green infrastructure accounted for 4% of the required capacity (McKinsey & Company, 2023). McKinsey and Company recommended giving private sector access to public land to develop green infrastructure, and this can be a suitable incentive (McKinsey & Company,

2023). Whether private sector should bear a greater burden in facilitating green growth is debatable.

The impacts of green growth in the short- and long term must be determined for justification of the required amount of investment (Fay, 2012). Overall green investment policies and strategies must be supported by available data, knowledge, and intelligence to build awareness and support for such initiatives. The role of the national statistical offices in the alignment of data-gathering initiatives about the SDGs, need to address specifically goals relating to land-based and water-based resources to ensure that green economy and blue economy frameworks can be effectively implemented. The responsible use of artificial intelligence to address the changing dynamics of resources on islands should be assessed. Elvira (2018) has recommended use of artificial intelligence in small island developing states vulnerable to climate change. The complexities and challenges of sustainable development can be managed by intelligence gathering, storage, and use.

The interrelationship between green investment and disaster risk reduction is still to be explored. A case for greater private sector investment in disaster risk reduction suggests that several strategies including the influence of the financial sector, development funding, certification schemes, and lessons learnt from previous disasters support building island resilience (Mahon, Becken, & Rennie, 2013). A lack of private sector investment in green technologies is a drawback to the sustainability of island destinations. More sources of financial capital are needed to build up resilient infrastructure in islands. Nowreen and Mohiuddin (2021) have suggested that the sustainability of financial capital is related to greater investment in locally owned businesses, green technology, and conservation. The sustainable development of small island developing states is affected by availability of funds to finance climate change adaptation. Given this funding challenge, a partnership between the public and private sector that supports climate change adaptation may be feasible. Hess and Kelman (2017) have proposed mechanisms initiated by the industry, namely water management and risk transfer, and mechanisms initiated by the government, namely taxes and levies, funds, and building codes and regulations, to support climate change adaptation.

13.4 Sustainable system transformation

A system transformation can occur with well-formulated and implemented policies. An island tourism policy framework to achieve sustainable development must consider the nature of the tourism sector on islands (Twining-Ward & Butler, 2002). Based on an island's attributes, tourism may take on various forms and characteristics, and enabling sustainability and resiliency constrained by the island's attributes. Scheyvens and Momsen (2008) have noted the forces that bring about change in small island states including colonialization and outside intervention, monetary economy, and physical constraints. The concept of being an islander and living in a village has changed and this affects sustainable system transformation in several ways. Island tourism policies may be formulated without community involvement if that sense of community has been lost. In examining three Pacific

Best Practice 2: Dominica and climate resiliency

In the aftermath of major hurricanes, the island of Dominica is working on climate resiliency in innovative ways. The geological character of the island of Dominica is based on the existence of several dormant volcanoes. The island is one of the few in the English-speaking Caribbean with an indigenous population, the Kalinago people. Hurricane Maria devastated the island in 2017 causing US$ 1.3 billion or 226% of the island's gross domestic product (World Bank, 2023). The government has established the Climate Resilience Execution Agency for Dominica (CREAD) (Jones, Doughorty, & Brown, 2022). The CREAD has developed a Climate Resilience and Recovery Plan 2020–2030 that stipulates three pillars: "climate resilient systems, disaster risk management and disaster response and recovery" (Climate Resilience Execution Agency for Dominica, 2020; Forster, Shelton, White, Dupeyron, & Mizinova, 2022, p. S69). A Disaster Vulnerability Reduction Project, funded by several international agencies, seeks to pilot adaptive measures, capacity building and data development, natural disaster response investment, and project management (Disaster Vulnerability Reduction Project, 2023). One of the targets of the resilience plan is to make communities to be self-sufficient for 2 weeks following a disaster (Wilkinson, 2023b). Within Dominica's sustainable future is a plan to become carbon neutral with investment in geothermal technology using its volcanic geology (Wilkinson, 2023b). Critical to Dominica's sustainability plans is the ongoing buy-in of stakeholders and education of communities about the plans (Wilkinson, 2023a).

Island countries, Stumpf and Swanger (2015) have found that Palau expressed a loss of control of their tourism industry because of external involvement. In that regard the capability of the island government to build capacity for policy-making, control elements of demand and supply, and support tourism development by appropriate planning approaches and tools are key for system transformation. Drawing on case studies in the Pacific Islands, Sofield (2003) has expressed the need for empowerment for sustainable tourism development.

From a system perspective, inputs and processes result in outputs and effective outcomes. System transformation means the system's outcome is a new entity and in the case of an island, the island transforms in a way that it is unrecognizable compared with its previous form. While system transformation may occur for radical transformation to occur, both physical and behavioural changes are necessary. Such behavioural change requires research to understand the enabling factors of change. Human intervention is one of the key ingredients in island sustainability transformation. Abson et al. (2017) have argued for transformational sustainability interventions through leverage points and have noted the need to target interactions between people and nature. Riechers, Balázsi, García-Llorente, and Loos (2021) have supported the idea of human–nature connectedness as a leverage point and have suggested that knowledge about appropriate interventions and the effects of interventions are needed for effective long-term sustainability transformation. Changing behaviour to achieve sustainable practices means re-educating a population and tourists alike in sustainability thinking. Kanagasapapathy (2023) has noted sustainable consumption behaviour in tourism and the need for re-education.

Another aspect of system transformation relates to the level of innovation. Innovation involves integrating new methods, new production systems, and new technologies (Schumpeter & Opie, 1951). The use of new methods and technologies must be prioritized to support sustainability in island destinations. Digitization of business processes, government administration, and activities in the tourism sector to control, measure, and monitor achievement of sustainability goals can support sustainable development. Hernández Sánchez and Oskam (2022) have noted the role of digital transformation in handling operational issues of small businesses in the Canary Islands during the pandemic. Digital transformation is necessary for the sustainable development of islands as tourism destinations. Spencer et al. (2023) have explored the concept of 'light green' and 'dark green' tourism strategies in relation to Caribbean tourism and have noted the need to embrace digital transformation as an urgent matter. Parra-López, Barrientos-Báez, and Sánchez (2023) have explored how to develop more resilient islands from the SDGs and have noted the importance of digital transformation as a process improvement strategy.

13.5 Sustainable and regenerative tourism

The idea of regeneration of island tourism activities is not new (Carlsen, 2015; Chapman & Speake, 2011), and tourism systems, sites, and realization of benefits require regeneration from time to time. Conceptually, regeneration of an island's tourism sector is a formidable task and ad hoc initiatives may not realize the intended outcomes. Dredge (2022) have proposed a tourism reinvention through the adoption of regenerative tourism and have recommended a shift in socio-ecological consciousness, transition to an adaptive system, and engagement with a bottom-up approach. Herein lies a wholistic approach to the concept of regenerative tourism. In essence, regenerative tourism means that a tourism system is structured, strategized, and operationalized with practices that allow for renewal (Parra-López et al., 2023). Renewal in essence benefits hosts and tourists alike as regenerative tourism practices align tourism experiences with natural and cultural systems (Ateljevic & Sheldon, 2022). Regenerative tourism is not unlike the concept of alternative tourism or Page's (2014) tourism amoeba growth in which the tourism cells replicate using sustainable resources. Some authors argue that regenerative tourism allows for the realization of sustainable, transformational tourism and growth (Bangwayo-Skeete & Skeete, 2021; Walker & Lee, 2022).

In islands with confined boundaries and constrained resources, regenerative mindfulness has several benefits. First, tourism's contribution to the economy will be broadened beyond the employment and foreign exchange rhetoric and focus on regenerative tourism experiences that align with an island's natural and cultural resources. Second, tourism practices will seek to refuse extraction and embrace renewal by immersion of existing resources in the creation of inclusive tourism experiences. Circular economy principles are critical for regenerative tourism practices (Schumann, 2020; Tomassini & Cavagnaro, 2022). Third, employment practices will seek to fairly reward service-providers for their contributions and will provide entrepreneurial opportunities for community members to gain benefits from the resources in island communities. Fourth, women will contribute to the tourism value chain in an equitable way. Females by their very nature can innovate the delivery of hospitable services by providing nurturing and caring effects on hospitality. Boluk and Panse (2022) have revealed the role of women entrepreneurs in the enhancement of community well-being through regenerative practices. Fifth, a wellspring of regenerative tourism practices in an island can result in systemic changes that shifts tourism from being a foreign-based system to being a local-based experiential, transformative industry that benefits both locals and tourists alike and thereby support sustainable development.

13.6 Conclusion

The term sustainable development future is an oxymoron since sustainability means that a future is inevitable, however, sustainability actions may not guarantee a future. In island destinations, natural processes seem greater than any effort a government may embark on to achieve sustainability. Islands such as Kiribati are purchasing land in Fiji as the future existence of the island itself is threatened by climate-related events (Reed, 2023). Herein lies the importance of rethinking the meaning of island tourism sustainability. Tourism as an economic activity can support island livelihoods, however, the focus must be on the sustainability of the island population rather than the sustainability of tourism. Readjustments in thinking about the contribution tourism makes to island life is important to fully consider a sustainable development future of island nations. In analyses of Barbados and Grenada, Bangwayo-Skeete and Skeete (2021) have noted transformations that contributed to sustainable tourism systems and modelled self-correcting forces and adjustments to build resilience. In large islands with diverse economic activities a sustainable development future is more realistic than in small islands that are dependent on tourism as the main economic growth activity. Bearing this in mind, sustainable development must be mapped out according to the characteristics of the island and understood by assessing the island's readiness for sustainable practices. Sustainable practices require policies and instruments that support sustainability and bringing about behavioural changes to enact sustainable and regenerative practices.

Chapter 13 discussion questions

1 Discuss the role of the Sustainable Development Goals in island tourism sustainability and resilience.
2 Evaluate the policies needed to facilitate green investment in an island destination.
3 Create a roadmap to achieve regenerative tourism in an island destination.

References

Abson, D. J., Fischer, J., Leventon, J., Newig, J., Schomerus, T., Vilsmaier, U., … Jager, N. W. (2017). Leverage points for sustainability transformation. *Ambio, 46*(June), 30–39.

Agenda, D. (2023). The Bridgetown Initiative: Here's everything you need to know. *World Economic Forum*. Retrieved from www.weforum.org/agenda/2023/01/barbados-bridget own-initiative-climate-change/

Agyeiwaah, E., McKercher, B., & Suntikul, W. (2017). Identifying core indicators of sustainable tourism: A path forward? *Tourism Management Perspectives, 24*(October), 26–33.

Altinay, L., Var, T., Hines, S., & Hussain, K. (2007). Barriers to sustainable tourism development in Jamaica. *Tourism Analysis, 12*(1–2), 1–13.

Ateljevic, I., & Sheldon, P. J. (2022). Guest editorial: Transformation and the regenerative future of tourism. *Journal of Tourism Futures, 8*(3), 266–268. doi:10.1108/JTF-09-2022-284

Bangwayo-Skeete, P. F., & Skeete, R. W. (2021). Modelling tourism resilience in small island states: A tale of two countries. *Tourism Geographies, 23*(3), 436–457.

Blancas, F. J., Lozano-Oyola, M., González, M., & Caballero, R. (2016). Sustainable tourism composite indicators: A dynamic evaluation to manage changes in sustainability. *Journal of Sustainable Tourism, 24*(10), 1403–1424.

Boluk, K. A., & Panse, G. (2022). Recognising the regenerative impacts of Canadian women tourism social entrepreneurs through a feminist ethic of care lens. *Journal of Tourism Futures, 8*(3), 352–366. doi:10.1108/JTF-11-2021-0253

Borgatti, S. P. (2002). *Netdraw network visualisation*. Harvard, MA: Analytic Technologies.

Borgatti, S. P., Everett, M. G., & Freeman, L. C. (2002). *Ucinet for Windows: Software for social network analysis*. Harvard, MA: Analytic Technologies.

Brohman, J. (1996). New directions in tourism for third world development. *Annals of Tourism Research, 23*(1), 48–70.

Burbano, D. V., Valdivieso, J. C., Izurieta, J. C., Meredith, T. C., & Ferri, D. Q. (2022). "Rethink and reset" tourism in the Galapagos Islands: Stakeholders' views on the sustainability of tourism development. *Annals of Tourism Research Empirical Insights, 3*(2), 100057. https://doi.org/10.1016/j.annale.2022.100057

Carlsen, J. (2015). Island tourism: Systems modelling for sustainability. In M. Hugues, D. Weaver, & C. Pforr (Eds.), *The practice of sustainable tourism* (pp. 83–94). Abingdon: Routledge.

Chapman, A., & Speake, J. (2011). Regeneration in a mass-tourism resort: The changing fortunes of Bugibba, Malta. *Tourism Management, 32*(3), 482–491.

Climate Resilience Execution Agency for Dominica (2020). *Dominica climate resilience and recovery plan 2020–2030*. Dominca: Climate Resilience Execution Agency for

Dominica. Retrieved from https://static1.squarespace.com/static/631647bc55d759246
4e828c2/t/635163b23466a51d14a90126/1666278342540/CRRP-Final-042020.pdf

Cole, S., & Razak, V. (2011). Island awash–sustainability indicators and social complexity in the Caribbean. In M. Budruk & R. Phillips (Eds.), *Quality-of-life community indicators for parks, recreation and tourism management* (pp. 141–161). London: Springer.

Connell, J. (2018). Islands: Balancing development and sustainability? *Environmental Conservation, 45*(2), 111–124.

Corfee-Morlot, J., Marchal, V., Kauffmann, C., Kennedy, C., Stewart, F., Kaminker, C., & Ang, G. (2012). Towards a green investment policy framework: The case of low-carbon, climate-resilient infrastructure. Retrieved from France: www.oecd-ilibrary.org/environm
ent/towards-a-green-investment-policy-framework_5k8zth7s6s6d-en

Disaster Vulnerability Reduction Project (2023). Project Description. Retrieved from https://
dvrp.gov.dm/project-description

Dredge, D. (2022). Regenerative tourism: Transforming mindsets, systems and practices. *Journal of Tourism Futures, 8*(3), 269–281.

Elvira, B. (2018). Deep reinforcement learning for intelligent road maintenance in small island developing states vulnerable to climate change: Using artificial intelligence to adapt communities to climate change. Retrieved from Sweden: https://uu.diva-portal.org/
smash/get/diva2:1278708/FULLTEXT03.pdf

Farsari, Y., & Prastacos, P. (2000). Sustainable tourism indicators: Pilot estimation for the municipality of Hersonissos, Crete. International Scientific Conference "Tourism on Islands and Specific Destinations" Chios 14-16 December 2000. Retrieved from Sustainable Tourism Indicators: Pilot Estimation for the Municipality of Hersonissos (diva-portal.org).

Fay, M. (2012). *Inclusive green growth: The pathway to sustainable development.* Washington, DC: World Bank Publications.

Forster, J., Shelton, C., White, C. S., Dupeyron, A., & Mizinova, A. (2022). Prioritising well-being and resilience to 'build back better': Insights from a Dominican small-scale fishing community. *Disasters, 46*(S1), S51–S77. doi:10.1111/disa.12541

Francis, H. (2012). Developing a self-sustaining protected area system: A feasibility study of national tourism fee and green infrastructure in the Solomon Islands. *Journal of Sustainable Finance & Investment, 2*(3–4), 287–302.

Hernández Sánchez, N., & Oskam, J. (2022). A "new tourism cycle" on the Canary Islands: scenarios for digital transformation and resilience of small and medium tourism enterprises. *Journal of Tourism Futures.* doi:10.1108/JTF-04-2022-0132

Hess, J., & Kelman, I. (2017). Tourism industry financing of climate change adaptation: Exploring the potential in small island developing states. *Climate, Disaster and Development Journal, 2*(2), 33–45.

Inter-American Development Bank (2018). Tourism strategy and action plan for Jamaica: Promoting resilience, sustainability, innovation & entrepreneurship. Retrieved from www.iadb.org/en/whats-our-impact/JA-T1149

Ivars-Baidal, J. A., Vera-Rebollo, J. F., Perles-Ribes, J., Femenia-Serra, F., & Celdrán-Bernabeu, M. A. (2021). Sustainable tourism indicators: What's new within the smart city/destination approach? *Journal of Sustainable Tourism, 31*(7), 1556–1582.

Jones, E., Dougherty, K., & Brown, P. (2022). 'Building back better' in the context of multi-hazards in the Caribbean. *Disasters, 46*(S1), S151–S165. doi:10.1111/disa.12545

Kanagasapapathy, D. (2023). How big is your tourism footprint?: In search of the sustainable tourist. In *Routledge handbook of trends and issues in tourism sustainability, planning and development, management, and technology* (pp. 69–80). Abingdon: Routledge.

Kelman, I. (2019). Critiques of island sustainability in tourism. *Tourism Geographies, 23*(3), 397–414.

Kennedy, V., Crawford, K. R., Main, G., Gauci, R., & Schembri, J. A. (2022). Stakeholder's (natural) hazard awareness and vulnerability of small island tourism destinations: A case study of Malta. *Tourism Recreation Research, 47*(2), 160–176.

Lee, T. H., & Hsieh, H.-P. (2016). Indicators of sustainable tourism: A case study from a Taiwan's wetland. *Ecological Indicators, 67*(August), 779–787.

Lim, C. C., & Cooper, C. (2009). Beyond sustainability: Optimising island tourism development. *International Journal of Tourism Research, 11*(1), 89–103.

Linnerooth-Bayer, J., Mechler, R., & Hochrainer, S. (2011). Insurance against losses from natural disasters in developing countries. Evidence, gaps and the way forward. *IDRiM Journal, 1*(1), 59–81.

Mahon, R., Becken, S., & Rennie, H. (2013). *Evaluating the business case for investment in the resilience of the tourism sector of small island developing states.* New Zealand: Lincoln University.

McElroy, J. L. (2003). Tourism development in small islands across the world. *Geografiska Annaler: Series B, Human Geography, 85*(4), 231–242.

McKinsey & Company (2023). Green infrastructure: Could public land unlock private investment? Retrieved from www.mckinsey.com/industries/public-sector/our-insights/green-infrastructure-could-public-land-unlock-private-investment

McLeod, M. (2023). Resilience building Caribbean tourism. In G. Sinclair-Maragh (Ed.), *The dynamics of Caribbean tourism, opportunities, challenges and a re-imagined future* (pp. 1–24).Jamaica: University of Technology, Jamaica Press.

Nair, V., & McLeod, M. (2020). Lessons learnt from the experience of countries in the Caribbean in aligning tourism investment, business and operations with the United Nations Sustainable Development Goals (SDGs). *Worldwide Hospitality and Tourism Themes, 12*(3), 353–358.

Nowreen, S., & Mohiuddin, M. (2021). The principles and practices of sustainable tourism investments and development in Bangladesh. In A. Hassan (Ed.), *Tourism in Bangladesh: Investment and development perspectives* (pp. 383–399). Singapore: Springer.

Page, S. J. (2014). *Tourism management.* Abingdon, UK: Routledge.

Parra-López, E., Barrientos-Báez, A., & Sánchez, M. d. l. Á. P. (2023). Island destinations in the face of global challenges. In A. Morrison & D. Buhalis (Eds.), *Routledge handbook of trends and issues in tourism sustainability, planning and development, management, and technology* (pp. 151–159). Abingdon, UK: Routledge.

Polnyotee, M., & Thadaniti, S. (2014). The survey of factors influencing sustainable tourism at Patong beach, Phuket Island, Thailand. *Mediterranean Journal of Social Sciences, 5*(9), 650–650.

Reed, B. (2023, February 23, 2021). Kiribati and China to develop former climate-refuge land in Fiji. Retrieved from www.theguardian.com/world/2021/feb/24/kiribati-and-china-to-develop-former-climate-refuge-land-in-fiji

Riechers, M., Balázsi, Á., García-Llorente, M., & Loos, J. (2021). Human–nature connectedness as leverage point. *Ecosystems and People, 17*(1), 215–221.

Sachs, J. D., Schmidt-Traub, G., Mazzucato, M., Messner, D., Nakicenovic, N., & Rockström, J. (2019). Six transformations to achieve the sustainable development goals. *Nature Sustainability, 2*(9), 805–814.

Scheyvens, R., & Momsen, J. (2008). Tourism in small island states: From vulnerability to strengths. *Journal of Sustainable Tourism, 16*(5), 491–510.

Schumann, F. R. (2020). Circular economy principles and small island tourism Guam's initiatives to transform from linear tourism to circular tourism. *Journal of Global Tourism Research, 5*(1), 13–20.

Schumpeter, J. A., & Opie, R. (1951). *The theory of economic development.* Cambridge: Harvard University Press.

Siedschlag, I., & Yan, W. (2021). Firms' green investments: What factors matter? *Journal of Cleaner Production, 310*(August), 127554.

Sofield, T. H. (2003). *Empowerment for sustainable tourism development.* Kidlington, UK: Pergamon.

Spencer, A. J., Lewis-Cameron, A., Roberts, S., Walker, T. B., Watson, B., & McBean, L. M. (2023). Post-independence challenges for Caribbean tourism development: A solution-driven approach through Agenda 2030. *Tourism Review, 78*(2), 580–613.

Stumpf, T., & Swanger, N. (2015). Tourism involvement-conformance theory: A grounded theory concerning the latent consequences of sustainable tourism policy shifts. *Journal of Sustainable Tourism, 23*(4), 618–637.

Tomassini, L., & Cavagnaro, E. (2022). Circular economy, circular regenerative processes, and placemaking for tourism future. *Journal of Tourism Futures, 8*(3), 342–345. doi:10.1108/JTF-01-2022-0004

Tsaur, S.-H., & Wang, C.-H. (2007). The evaluation of sustainable tourism development by analytic hierarchy process and fuzzy set theory: An empirical study on the Green Island in Taiwan. *Asia Pacific Journal of Tourism Research, 12*(2), 127–145.

Twining-Ward, L., & Butler, R. (2002). Implementing STD on a small island: Development and use of sustainable tourism development indicators in Samoa. *Journal of Sustainable Tourism, 10*(5), 363–387.

United Nations Environment Programme (2023). Facilitating green investment and financing. Retrieved from www.unep.org/explore-topics/green-economy/what-we-do/economic-and-fiscal-policy/fiscal-policy/policy-analysis-0

UNWTO (2004). *Indicators of sustainable development for tourism destinations.* Madrid: World Tourism Organization.

Walker, T. B., & Lee, T. J. (2022). Contributions to sustainable tourism in small islands: An analysis of the Cittàslow movement. In M. McLeod, R. Dodds, & R. Butler (Eds.), *Island tourism sustainability and resiliency* (pp. 54–74). Abingdon: Routledge.

Wilkinson, E. (2023a). How a small Caribbean island is trying to become hurricane-proof. *The Conversation.* Retrieved from https://theconversation.com/how-a-small-caribbean-island-is-trying-to-become-hurricane-proof-217999

Wilkinson, E. (2023b). How is Dominica trying to become the world's first climate-resilient nation? Retrieved from www.weforum.org/agenda/2023/11/dominica-climate-resilient-nation/

World Bank (2023). Dominica's journey to become the world's first climate resilient country. Retrieved from www.worldbank.org/en/news/feature/2023/09/26/dominica-s-journey-to-become-the-world-s-first-climate-resilient-country

World Economic Forum (2013). The green investment report: The ways and means to unlock private finance for green growth. Retrieved from Geneva, Switzerland: www3.weforum.org/docs/WEF_GreenInvestment_Report_2013.pdf

14 Island tourism prospects

14.1 Introduction

The purpose of this book is to explore policies, governance, and planning in the sustainable development of the tourism industry on islands. Islands are unique features with different characteristics that shape the offerings of tourism products and services. Islands are small or large landscapes with small or large populations, and these differences affect the scale, nature, and sustainability of tourism activities on islands (Choi & Sirakaya, 2010; Weaver, 2017). In large islands, economic activities around tourism may be limited and domestic tourism is a likely development path. In small islands, dependence on international tourism is a likely path for economic growth. Barrowclough (2007) has noted the difference between tourism investment in large and small islands as large islands have greater knowledge and management capacity to support franchising. This distinction is important for addressing the range of policies that guide and steer sustainable tourism development. Islands are at different stages of tourism development (Butler, 1980), and this affects island tourism sustainability.

The extent to which tourism development occurs on islands has been modelled (Butler, 1980; McElroy & De Albuquerque, 1998); however, island tourism sustainability requires elaboration. This volume has discussed the phenomenal challenges regarding climate change impacts and resiliency. Building a resilient island tourism destination starts with understanding the range of internal resources that form the foundation of defending some of the vulnerabilities islands cope with. Baldacchino (2015) has pointed out the gamut of economic vulnerabilities of small islands and has noted the threat of emigration and the need for economic resiliency. Building resiliency also must be future thinking as new planning methodologies have to be delivered to address the dynamics of the global and natural forces affecting islands. Island tourism recovery is complicated by the destruction of critical tourism infrastructure at air and sea ports (McLeod, 2022). This volume highlighted specific case studies across the globe to create learning paths for policymakers and researchers, and to address the need for more research about the challenges islands are facing as tourism-led growth slows down or reverses.

DOI: 10.4324/9781003435112-17

On the matter of sustainable development this volume explores sustainability indicators to reveal the more important indicators and the interrelationships of the indicators with the Sustainable Development Goals (SDGs). Ng, Chia, Ho, and Ramachandran (2017) have selected 9 sustainability indicators based on the specific context in examining ecotourism sustainability on Tioman Island, Malaysia. Lists of island tourism sustainability indicators are endless. Nonetheless, sustainability indicators, both positive and negative, are resources that result in system growth or decline (Font et al., 2021; Torres-Delgado & Palomeque, 2018; Walker & Lee, 2022). Kurniawan, Adrianto, Bengen, and Prasetyo (2019) have used a social–ecological approach to evaluate the status of sustainable development in small islands and have suggested that indicators improve the results. The tourism system in islands is affected by several forces, and understanding the future of tourism in islands relates to unpacking the sustainability indicators (Font et al., 2021). Policies for sustainable development must focus on addressing the more important indicators of island tourism sustainability (Farsari, 2021; Wolf et al., 2022).

14.2 Sustainable tourism development policy

The first part of this volume discussed island tourism policies. Tourism policies are needed to guide tourism growth and sustainable development. In the Caribbean, tourism industries have been faced with ongoing global threats and shocks that warrant timely tourism policy-making to adjust and adapt to new global realities. Policies must be supported by easily accessible data and information that facilitate knowledge creation and innovative capacity. Policies relating to demand were explored. In the context of sustainability, islands must set limits of tourism demand going forward (Ministry of Tourism French Polynesia, 2022). Tourism is not a mass game but an exclusive game when it comes to island tourism sustainability. Island governments must be clear about how tourism activities affect island resources and the future of the tourism system. Ridderstaat and Nijkamp (2016) have noted the vulnerability of small island destinations based on attracting visitors from fewer markets of origin, and an island's distance from these markets is another vulnerability. Bearing exclusivity in mind, islands must make choices about the tourism markets to attract, and the specific tourists from those markets. Working with specialized intermediaries, island tourism destinations need to become specific, specialized, and selective about the nature of the tourism industry on island.

Specific, specialized, and selective policies to address tourism demand and supply were detailed in this volume. Policies relating to airlift, marketing, and visa restrictions are important to facilitate demand for island tourism. Destination branding as a specific policy must also be addressed as islands transform imagery that will depict sustainability principles. Supply-led policies that allow locals to participate in the tourism value chain are also important. Entrepreneurship and a greater role of micro-, small-, and medium-sized enterprises in tourism must be a priority for sustainable development. Innovation must be built on both the demand and supply sides of island tourism policies. The capacity building for policymaking, formulating, implementing, and taking must be an approach to create innovative outcomes in island tourism policy development processes.

14.3 Sustainable tourism development governance

Stakeholder involvement is a key activity for the success of sustainable development on islands. Governance of island tourism must be viewed from a metagovernance perspective that includes public sector, private sector, civil society, and community actors working together for a common purpose. McLeod (2023) has illustrated the four roles of tourism stakeholders involved in policy networks. Governance actors must be accountable and responsible for the adoption of sustainability principles. In that regard, building awareness, education, and training capacity within island destinations improve knowledge about sustainability. Shared knowledge builds a collaborative environment as stakeholders learn to support activities for sustainable development. Consensus building is also supported by knowledge sharing, and within large island destinations a greater effort is needed to build consensus around specific plans for sustainable tourism development.

Herein is the need to understand network governance in island tourism activities. First, network governance seeks to involve all stakeholders in some way in tourism policy development activities. The interrelationships of the stakeholders influence decisions about tourism sustainability. Second, network governance must be built out within a governance framework that supports sustainability. The two sustainable development frameworks discussed in this volume are the green economy and the blue economy. Sustainable tourism development on islands must be supported by a framework that combines tourism activities in an ecosystem wherein the interactions between land-based activities and marine-based activities converge and an integrated framework was proposed. An island ecosystem is very sensitive to human activities, and tourism activities use an island's 'free' resources to a greater extent than other industries. As a result, the governance of island tourism must be delivered using an ecosystem perspective that handles all activities in a wholistic manner. Third, governance must be viewed using the environmental, social, and governance (ESG) perspective combined with political and economic strategies as well. ESG has been used in assessing investors' contributions to development, however, such a framework can be broadly adopted within all levels of governance activities. Finally, the concept of sustainable governance (Stiftung, 2011) that expounds democratic principles, policy performance, and executing agency capacity and accountability has been discussed in relation to sustainable development of island tourism.

14.4 Sustainable tourism development planning

Planning is a critical activity for sustainability. Integrated planning that addresses the myriad of developmental aspects of island tourism must be given priority. The master planning approach has been conducted in several islands (Commonwealth of Dominica, 2013; Government of the Republic of Maldives, 2023); however, the issue is with the implementation of a master plan (Teniwut, Hamid, & Makailipessy, 2022; Wolf et al., 2022). As such, island tourism planning activities require new approaches to make tourism plans workable. Outcome mapping methodology and scenario planning provide new possibilities for instituting and implementing sustainable development practices on islands. One of the drawbacks that affects the implementation of tourism plans is the lack of involvement of local stakeholders (Dodds & Butler, 2009).

A practical approach for assessing sustainability using indicators and the SDGs in island tourism was detailed in this volume. Such an approach can be adopted to ensure that efforts towards sustainability are feasible and that the SDGs are achieved. The competing resources on islands, particularly when island governments decide to diversify the economy (Hampton & Christensen, 2007; Hess & Kelman, 2017), makes planning for sustainability difficult and complicated. As a result, along with an assessment of the sustainability indicators must be a clear plan for tourism-led growth in an island economy and any resultant effects such as environmental degradation, as suggested by Fauzel and Tandrayen-Ragoobur (2023). Keeping tourism plans on track requires government commitment, accountability, and transparency. Such commitment must also be supported within a global environment and the actions of development agencies.

14.5 Conclusion

Islands have smaller populations than large metropolitan areas, and the ability to command and negotiate preferential terms and conditions is considerably less. As a result, a greater voice for sustainable development must be articulated for more attention to be given to the needs of island nations and countries. Placing islands on a global stage means that policies that steer island sustainable development will gain greater traction. Tourism as an activity on islands has been written about over several decades, and yet a sustainable tourism model for islands must be framed given the myriad of political, historical, cultural, natural, and economic contexts that island landscapes are founded on. Cheer (2020) has articulated the urgency for sustainability indicators in small islands. Tourism can place islands on a sustainable development path with well-formulated and managed policies within a governance and planning framework for sustainable development. Bearing this in mind, this volume has addressed policies, governance, and planning to provide island tourism destinations with appropriate tools to consider and support sustainable development. Sustainability and resiliency of island tourism are a must for the benefit of island populations globally.

References

Baldacchino, G. (2015). Small island states and territories: Vulnerable, resilient, but also doggedly perseverant and cleverly opportunistic. In G. Baldacchino (Ed.), *Entrepreneurship in small island states and territories* (pp. 1–28). Abingdon: Routledge.

Barrowclough, D. (2007). Foreign investment in tourism and small island developing states. *Tourism Economics, 13*(4), 615–638.

Butler, R. W. (1980). The concept of a tourist area cycle of evolution: Implications for management of resources. *Canadian Geographer/Le Géographe canadien, 24*(1), 5–12.

Cheer, M. (2020). The urgency for sustainability indicators. *Institute of Island Studies.* Retrieved from https://islandstudies.com/files/2022/08/Annual-Report-on-Global-Islands-2019-Chapter-5-Tourism-on-small-islands-The-urgency-for-sustainability-indicators-Joseph-M.-Cheer.pdf

Choi, H. C., & Sirakaya, E. (2010). Sustainability indicators for managing community tourism. In M. Budruk & R. Phillips (Eds.), *Quality-of-life community indicators for parks, recreation and tourism management* (Vol. 43, pp. 115–140). London: Springer.

Commonwealth of Dominica (2013). Tourism master plan 2012–2022. Dominica: Government of the Commonwealth of Dominica. Retrieved from https://tourism.gov.dm/images/documents/tourism_master_plan/tourism_master_plan_june2013.pdf

Dodds, R., & Butler, R. (2009). Barriers to implementing sustainable tourism policy in mass tourism destinations. *Tourismos: An International Multidisciplinary Journal of Tourism, 5*(1), 35–53.

Farsari, I. (2021). Exploring the nexus between sustainable tourism governance, resilience and complexity research. *Tourism Recreation Research, 48*(3), 352–367.

Fauzel, S., & Tandrayen-Ragoobur, V. (2023). Sustainable development and tourism growth in an island economy: A dynamic investigation. *Journal of Policy Research in Tourism, Leisure and Events, 15*(4), 502–512.

Font, X., Torres-Delgado, A., Crabolu, G., Palomo Martinez, J., Kantenbacher, J., & Miller, G. (2021). The impact of sustainable tourism indicators on destination competitiveness: The European Tourism Indicator System. *Journal of Sustainable Tourism, 31*(7), 1608–1630.

Government of the Republic of Maldives (2023). Maldives fifth tourism master plan 2023–2027: Goals and strategies. Republic of Maldives. Retrieved from www.tourism.gov.mv/dms/document/4969b4831928f1bdf3506340fb6974fc.pdf

Hampton, M. P., & Christensen, J. (2007). Competing industries in islands a new tourism approach. *Annals of Tourism Research, 34*(4), 998–1020.

Hess, J., & Kelman, I. (2017). Tourism industry financing of climate change adaptation: Exploring the potential in small island developing states. *Climate, Disaster and Development Journal, 2*(2), 33–45.

Kurniawan, F., Adrianto, L., Bengen, D. G., & Prasetyo, L. B. (2019). The social–ecological status of small islands: An evaluation of island tourism destination management in Indonesia. *Tourism Management Perspectives, 31*(July), 136–144.

McElroy, J. L., & De Albuquerque, K. (1998). Tourism penetration index in small Caribbean islands. *Annals of Tourism Research, 25*(1), 145–168.

McLeod, M. (2022). Tourism destination recovery, a case study of Grand Bahama Island. In I. Bethell-Bennett, S. Rolle, J. Minnis, & F. Okumus (Eds.), *Pandemics, disasters, sustainability, tourism* (pp. 93–108). Leeds, UK: Emerald Publishing.

McLeod, M. (2023). Tourism policy networks in four Caribbean countries. *Annals of Tourism Research Empirical Insights, 4*(2), 100113.

Ministry of Tourism French Polynesia (2022). *Fāri'ira'a Manihini 2027: The welcome that reflects us and binds us together, tourism development strategy for French Polynesia 2022–2027*. Tahiti: Ministry of Tourism French Polynesia. Retrieved from www.calameo.com/read/00346150392c0b997b7c5

Ng, S. I., Chia, K. W., Ho, J. A., & Ramachandran, S. (2017). Seeking tourism sustainability – A case study of Tioman Island, Malaysia. *Tourism Management, 58*(February), 101–107.

Ridderstaat, J. R., & Nijkamp, P. (2016). Small island destinations and international tourism: Market concentration and distance vulnerabilities. In M. Ishihara, E. Hoshino, & Y. Fujita (Eds.), *Self-determinable development of small islands* (pp. 159–178). Singapore: Springer.

Stiftung, B. (2011). Sustainable governance indicators 2011: Policy performance and governance capacities in the OECD. Retrieved from Germany: https://api.pageplace.de/prev iew/DT0400.9783867933933_A18798354/preview-9783867933933_A18798354.pdf

Teniwut, W. A., Hamid, S. K., & Makailipessy, M. M. (2022). Developing a masterplan for a sustainable marine sector in a small islands region: Integrated MCE spatial analysis for decision making. *Land Use Policy, 122*(November), 106356.

Torres-Delgado, A., & Palomeque, F. L. (2018). The ISOST index: A tool for studying sustainable tourism. *Journal of Destination Marketing & Management, 8*(June), 281–289.

Walker, T. B., & Lee, T. J. (2022). Contributions to sustainable tourism in small islands: an analysis of the Cittàslow movement. In M. McLeod, R. Dodds, & R. Butler (Eds.), *Island tourism sustainability and resiliency* (pp. 54–74). Abingdon: Routledge.

Weaver, D. B. (2017). Core–periphery relationships and the sustainability paradox of small island tourism. *Tourism Recreation Research, 42*(1), 11–21.

Wolf, F., Moncada, S., Surroop, D., Shah, K. U., Raghoo, P., Scherle, N., … Havea, P. H. (2022). Small island developing states, tourism and climate change. *Journal of Sustainable Tourism*, 1–19. doi:10.1080/09669582.2022.2112203

Index

For Product Safety Concerns and Information please contact our EU
representative GPSR@taylorandfrancis.com
Taylor & Francis Verlag GmbH, Kaufingerstraße 24, 80331 München, Germany

www.ingramcontent.com/pod-product-compliance
Lightning Source LLC
Chambersburg PA
CBHW060308220326
41598CB00027B/4271

* 9 7 8 1 0 3 2 5 6 3 6 0 2 *